Cucumber Economic Values and Its Cultivation and Breeding

Edited by Haiping Wang

Published in London, United Kingdom

IntechOpen

Supporting open minds since 2005

Cucumber Economic Values and Its Cultivation and Breeding
http://dx.doi.org/10.5772/intechopen.87508
Edited by Haiping Wang

Contributors
Maadh Abdulwahab Alfahad, Gograj Singh Jat, Tusar Behera, Suman lata, Sachin Kumar, Olumide Samuel Daramola, Shampa Chakraborty, Sadhana Rayalu, Ravi Mohan Srivastava, Chidiebere Ugwu, Stephen Suru, Iyabo Bosede Adeoye, Sarmad Ghazi Al-Shawi, Sadiq Jaafir Aziz Alneamah, Oluwayemisi Ibitoye, Taofeek Ajiboye, Azeemat Titilola Abdulazeez, Rukayat Abiodun Oyegoke, Hamdalat Folake Muritala, Fatimah Aluko Abubakar, Bernard Ndubuisi Okafor, Japhet J. Yaduma, Olawale Arogundade, Titilayo Ajose, Joshua Matthew, Itunu Osijo, Michael Henry Henry Böhme, Haiping Wang, Huixia Jia, Sneha Joshih

First published in London, United Kingdom, 2021 by IntechOpen
IntechOpen is the global imprint of INTECHOPEN LIMITED, registered in England and Wales, registration number: 11086078, 5 Princes Gate Court, London, SW7 2QJ, United Kingdom
Printed in Croatia

British Library Cataloguing-in-Publication Data
A catalogue record for this book is available from the British Library

Additional hard and PDF copies can be obtained from orders@intechopen.com

Cucumber Economic Values and Its Cultivation and Breeding
Edited by Haiping Wang
p. cm.
Print ISBN 978-1-83968-023-6
Online ISBN 978-1-83968-024-3
eBook (PDF) ISBN 978-1-83968-025-0

For EU product safety concerns:
IN TECH d.o.o., Prolaz Marije Krucifikse Kozulić 3, 51000 Rijeka, Croatia, info@intechopen.com or visit our website at intechopen.com.

Meet the editor

Haiping Wang is a full-time research scientist and professor in the Department of Vegetables Germplasm, Institute of Vegetables and Flowers, Chinese Academy of Agricultural Sciences (IVFCAAS). His research interests include vegetable genetic resources and preservation of the diversity of Midterm Gene-Bank of Vegetables Genetic Resources in China. Research on vegetable genetics is conducted to improve the crop for growers and consumers. His key areas of interest include garlic, ginger, radish, and cucumber genetics and the development of genomic tools. His outreach activities include interaction with the garlic and ginger production and with consumers. Dr. Wang is the author and co-author of seventy publications in scientific journals and thirteen book chapters in Chinese and English. He has reviewed numerous publications for more than ten international scientific journals.

Contents

Preface XI

Section 1
Introduction 1

Chapter 1 3
Introductory Chapter: Studies on Cucumber
by Huixia Jia and Haiping Wang

Section 2
Economic Values and Pharmacological Potentials 11

Chapter 2 13
Antidiabetic Principle in *Cucumis sativus* L.
by Ibitoye Oluwayemisi Beatrice, Ajiboye Taofeek Olakunle, Abdulazeez Azeemat Titilola, Oyegoke Rukayat Abiodun, Muritala Hamdalat Folake and Abubakar Fatimah Aluko

Chapter 3 23
Cucumber Pickles and Fermentations
by Sarmad Ghazi Al-Shawi and Sadiq Jaafir Aziz Alneamah

Chapter 4 39
Cosmetic, Culinary and Therapeutic Uses of Cucumber (*Cucumis sativus* L.)
by Chidiebere Ugwu and Stephen Suru

Section 3
Production and Market 49

Chapter 5 51
Market Performance and Structure of Cucumber in Ibadan, Oyo State
by Iyabo Bosede Adeoye

Section 4
Cultivation and Management 63

Chapter 6 65
Growth Promoting and Stabilizing of Cucumber Plants Cultivated in Soilless Cultivation Systems Using Biostimulators
by Michael Henry Böhme

Chapter 7 89
Soil and Agronomic Management for Cucumber Production in Nigeria
by Bernard Ndubuisi Okafor and Japhet J. Yaduma

Chapter 8 99
Health Beneficial Effects of Cucumber
by Shampa Chakraborty and Sadhana Rayalu

Chapter 9 109
Weed Interference and Management in Cucumber (*Cucumis sativus* L.)
by Olumide Samuel Daramola

Section 5
Pest and Disease Prevention 127

Chapter 10 129
Integrated Pest Management for Cucurbits in Cucumber (*Cucumis sativus* L.)
by Ravi Mohan Srivastava and Sneha Joshih

Chapter 11 145
The Activity of New Bio-Agent to Control *Cucumovirus Cucumber Mosaic Virus* (CMV)
by Maadh Abdulwahab Alfahad

Chapter 12 179
Current and Emerging Pests and Diseases of Cucumber (*Cucumis sativus* L.) in Africa
by Olawale Arogundade, Titilayo Ajose, Joshua Olumide Matthew and Itunu Adeyemi Osijo

Section 6
Breeding Progress 199

Chapter 13 201
Classical Genetics and Traditional Breeding in Cucumber (*Cucumis sativus* L.)
by Gograj Singh Jat, Tusar Kanti Behera, Suman Lata and Sachin Kumar

Preface

Cucumber is an economically important fruit vegetable. With abundant water, nutrients and phytochemical composition, cucumber has versatile uses in culinary, therapeutic and cosmetic purposes. Cucumber has also multiple advantages such as being a diploid and having a small genome, short life cycle and self-compatible mating system, so it has been identified as a model plant for genetic studies. The scientific research of cucumber will be beneficial for its efficient protection and utilization.

Cucumber was cultivated about 3000 years ago. It is indigenous to India and likely originated from the foothills of the Himalayas, and spread to Western Asia and Southern Europe. Cucumber was introduced respectively to North China through the Silk Route and to South China from Burma and the India–China border, and subsequently spread to East Asia. Nowadays, cucumber is widely cultivated in temperate and tropical regions throughout the world.

Over six sections, this book presents the latest findings and research on several aspects of cucumber biology, breeding, cultivation, marketing, and more. Section 1, "Introduction," provides an overview of cucumber. Section 2, "Economic Values and Pharmacological Potentials," helps readers to understand the nutrients, functional components, and extensive utilization of cucumber. Section 3, "Production and Market," explains the market structure and earning situation related to selling cucumber. Section 4 "Cultivation and Management" and Section 5 "Pest and Disease Prevention" discusses the cultivation and improvement of cucumber to obtain high yield. Section 6, "Breeding Progress," summarizes the classical genetics and traditional breeding of cucumber, as well as presents prospects for future breeding.

I would like to extend thanks to all authors who contributed to this book. My sincere thanks go also Author Service Manager Ms. Jasna Bozic at IntechOpen who helped me through all the stages of preparing and publishing this book.

Haiping Wang
Institute of Vegetables and Flowers,
Chinese Academy of Agricultural Sciences,
Beijing, China

Section 1

Introduction

Chapter 1

Introductory Chapter: Studies on Cucumber

Huixia Jia and Haiping Wang

1. Introduction

Cucumber (*Cucumis sativus* L.) belongs to *Cucumis* genus in *Cucurbitaceae* family and is an economically important fruit vegetable. There are three wild or semi-wild varieties of cucumber: *C. sativus* L. var. hardwickii, *C. sativus* L. var. sikkimensis, *C. sativus* L. var. xishuangbannanesis. Cucumber is indigenous to India and likely originated from the foothills of the Himalayan Mountain [1, 2]. Cucumber was cultivated ~3000 years ago in India, and it seems to spread rapidly to Western Asia, and then to Southern Europe [2]. Cucumber was introduced respectively to North China through the Silk Route and to South China from Burma and India-China border, and subsequently spread to East Asia [2]. Genome variation analysis showed cucumber core germplasms were divided into four geographic groups including India, Eurasia, East Asia, and Xishuangbanna [3]. Nowadays, cucumber is widely cultivated in temperate and tropical regions throughout the world [4]. The total production of cucumber was 87,805,086 tons worldwide, and Asia was the largest producer accounting for 84.9% of the world's total production in 2019 (www.fao.org/faostat/en/). With abundant water, nutrients and phytochemical composition, cucumber has versatile uses in culinary, therapeutic and cosmetic purposes [5, 6]. Cucumber has multiple advantages such as diploid, small genome, short life cycle and self-compatible mating system, so it is suitable for genetic studies. Moreover, cucumber has been identified as a model plant for studying sex determination and plant vascular biology [7]. Consequently, numerous studies have been conducted to discover the miracle of cucumber. The book will cover the extensive benefits, production and market, cultivation and management, pests and diseases, breeding progress of cucumber.

2. Biological characteristics

Cucumber is an annual climbing herbaceous plant. The root system is shallow and mainly distributes in the cultivated land layer of 30 cm. The stem is vine with different degree of apical dominance. The cross section of the stem is rhombus, and the epidermis of the stem has burrs. The axillae on the stem have the ability of branching, and the number of branching varies greatly among varieties. The cotyledons of cucumber are opposite and long elliptic; euphylla are alternate, simple, pentagonal palmate or cordate in outline, and the blades are 3–7 lobed. The flower is axillary, unisexual and occasionally hermaphrodite. The calyx is green with bristles, and the corolla is yellow. The colour of young fruit changes from white to pale green, while mature fruit is yellow or brown when ripened. The shape of the fruit is diverse, such as clublike, cylindrical and spherical. Each fruit has 100–400 seeds. The weight of 1000 seeds is about 20–40 g.

3. Culinary, therapeutic and cosmetic uses of cucumber

At present, cucumber is the fourth most widely cultivated vegetable after tomato, cabbage and onion [8]. Cucumber has versatile uses in culinary, therapeutic and cosmetic purposes [5, 6]. Nutritional and epidemiological researches have shown various benefits of cucumber. For example, cucumber contains abundant nutrients and has crunchy texture and unique flavor, so it is a quintessential vegetable used for a variety of dishes, and it is also indispensable for salad, soup and smoothie. Cucumber is rich in superior hydration and phytochemicals, which have diverse health benefits including weight loss, anti-inflammation, remedy for multiple diseases of eczema, constipation, hypertension, atherosclerosis, cancer, etc. [9]. Recent studies found that the presence of kaempferol in cucumber is an important antidiabetic agent [10]. Furthermore, cucumber is popularly used for natural beautification and for skin treatments [11].

4. Influence factors and solutions of cucumber fermentation

Cucumber pickles are most commonly fermented vegetable and widely consumed throughout the world [12]. Good fermentation depends on the proper combinations and interactions of multiple physical, chemical and microbiological factors [13]. Brine storage and process operations are susceptible to oxidation reactions during the fermentation process, and this has adverse influence on the quality property of cucumber pickles. To control the influence factors of cucumber fermentation, researchers have done many efforts on modern and advanced technologies, such as reducing the concentration of brining sodium chloride, developing the brining properties using lactic acid bacteria cultures, developing an anaerobic tank system, preventing cucumber gaseous deterioration by pouring of CO2 from fermentation brines [13]. After storing the brine, excess salt and organic wastes need to be leached to complete the product processing, and these wastes are sources of serious environmental concern. Thus, the waste disposal needs to be solved in the cucumber pickle industry.

5. Performance, structure and constraints of cucumber market

Marketing is vital for linking production and consumption and facilitating agricultural productivity and employment [14]. Market performance is the ultimate result of various market activities, and market structure is the organization characteristics of the market that influence the nature of competition and pricing [15]. Both male and female participate in cucumber marketing, and the male–female rate has great differences in different regions. The wholesalers are older than the retailers. In Ibadan, most of the retailers were within 31–40 years age, whereas most of the wholesalers were within 41–50 years age. It's gratifying that cucumber marketing is usually profitable for the retailers and wholesalers at both peak and lean seasons of cucumber production. However, the cucumber market is competitive, and inequality exists in the market. Commodity perishability is an important constraint in cucumber market. Thus, it is indispensable to reduce perishable degree and prolong storage time after post-harvest.

6. Soil moisture and fertilizer management of cucumber

Inappropriate farming systems and poor agronomic management are responsible for low yield of cucumber. The quality/fertility status of soils is essential for growth and development for cucumber [16]. With good moisture and fertilizer management, optimum yield of cucumber might be attained. The conventional irrigation methods including flooding irrigation, furrow irrigation and drip irrigation have been widely applied for a long time in cucumber cultivation because of their low cost or simple operation [17, 18]. However, these irrigation methods are surface irrigation and are driven by positive pressure, which may cause low water use efficiency, water wastage and nutrient loss [16, 19]. To solve these problems, new irrigation technique such as negative pressure irrigation that controls automatically water release based on the soil water potential difference should be encouraged [16]. Inadequate fertilizer use causes low soil fertility that cannot provide sufficient nutrients for the normal growth of cucumber. The integration application of inorganic and organic fertilizer is more beneficial than the sole use of inorganic fertilizer or organic manures in cucumber production [20]. Moreover, fertilizer sources need to dissolve or decompose to make nutrients available for cucumber plants, so soil fertility also depends on soil water, temperature and density. Consequently, the soil management strategies such as negative pressure irrigation, seasonable fertilization, application of organic mulches and conservation tillage should be appropriately applied for sustainable production of cucumber.

7. Biostimulators promote growth of cucumber under soilless cultivated condition

Soilless cultivation in substrate culture is an important cultivation pattern for cucumber in greenhouses. The substrates should have specific physical properties including pore volume, air and water capacity, and density of substrates. Studies indicate that biostimulators can stabilize the production process to enhance plant growth under stress conditions. For instance, humate can increase vitality and growth of plants, improve seed germination, promote nutrient uptake, enhance transport and availability of micronutrients, and increase ion-exchange capacity. Lactates can produce bioregulatory effects to improve nutrient balance and plant vitality [21]. *Bacillus subtilis*, as a microorganism from the rhizosphere, can accelerate plant growth, stimulate the process of formation of plant organs, and enhance the resistance of biotic and abiotic stresses [21]. Application of biostimulators mixture (humate, lactate, and *Bacillus subtilis*) prevent growth reduction of cucumber under pH and temperature stresses through enhancing the root growth, whereas the growth is markedly reduced under stresses if no biostimulator is applied.

8. Pests and diseases during cucumber cultivation and production

During growth process, cucumber might be affected by multiple insect pests and diseases, resulting in decrease of yield and quality. The major insect pests in cucumber including *Diabrotica undecimpunctata*, *Acalymma vitatum*, *Bactrocera cucurbitae*, *Raphidopalpa foveicollis*, *Epilachna implicate*, *Myzus persicae*, *Aphis gossypii*, *Anasa tristis*, *Trialeurodes vapurariorus*, *Bemisia tabaci* and *B. argentifolii* [22, 23]. Currently, the pest management mainly relies on chemical pesticides that cause environmental pollution, pest resistance, and disturbance of balance between the pests and natural enemies. Moreover, this control strategy is harmful to human health. Therefore, an integrated pest management including pest monitoring, cultural method, host resistance, botanicals, biological control, and judicious use of chemicals is recommended for controlling these pests [24, 25] Many diseases caused by viral, bacterial, fungal and nematode pathogens severely affect the cultivation and production of cucumber. Viruses infecting cucumber belong to three genera: *Potyvirus*, *Cucumovirus* and *Crinivirus* [26]. Especially, the CMV, ZYMV, WMV, MWMV, PRSV and BPYV are major viruses that cause severe symptoms to cucumber. Downy mildew, powdery mildew and anthracnose also cause substantial losses of cucumber production [27]. Some pathogenic fungi including *Alternaria tenuis*, *Fusarium equisett*, *Phytophthora capsici*, *Botrytis cinerea* and *Cladosporium tenuissimum* cause rotting and high post-harvest losses of cucumber [28]. Furthermore, root-knot nematodes are prevalent destructive pathogens of cucumber [29]. Though a series of chemicals have been evaluated and screened to control these diseases, the biological control strategy and high-resistant varieties of cucumber need be developed and created to resist diseases in efficient and environmental ways.

9. Polyphenols act as antioxidants in cucumber to defense stresses

Plant secondary metabolites play important roles in adapting to various environments and defensing against biotic and abiotic stresses. Cucumber is a rich source of phenolic compounds that are important secondary metabolites [30, 31]. The antioxidant capacity of cucumber seems to be attributing to polyphenols that scavenge singlet oxygen, hydroxyl and lipid peroxyl radicals to prevent lipid oxidation. Better understanding of the molecular regulation of polyphenols biosynthesis is crucial to increase the production of polyphenols. Polyphenols are derivatives of phenylpropanoid pathway which involves an array of enzymes. Among these, phenylalanine ammonia lyase, chalcone synthase, cinnamate 4-hydroxylase and dihydroflavonol reductase play important roles [32]. In-depth study of these key enzymes in cucumber will facilitate to reveal the molecular mechanism of polyphenol synthesis, which is helpful for advancement in biotechnological and industrial applications.

10. Progress of traditional breeding and molecular breeding in cucumber

In the past decades, traditional breeding has played essential roles in cultivar innovation of cucumber. Some superior varieties with early maturity, high yield and high resistance have been developed through hybridization and mutagenesis [33]. However, this progress is slow because of the long cycle and difficulty in selection of stable genetic characters or genotypes. To overcome the obstacle of traditional breeding, molecular breeding technologies including molecular marker assisted breeding, genome-wide design breeding and genetic engineering have been applied in cucumber to accelerate the breeding cycle and select desirable traits. Molecular breeding of cucumber has made some progress and achievements on completion of genomics, genetic architecture and molecular mechanism underlying important traits, and creation of high quality and multi-resistant varieties [7, 34–36]. With increasing consumption demand of cucumber, more new varieties with excellent comprehensive properties are in need, and we might make some efforts from the following aspects: (i) expanding collection and utilization of cucumber germplasm resources; (ii) establishing highly efficient gene editing and genetic transformation technologies in cucumber; (iii) identifying new loci or genes associated with key agronomic traits of cucumber and combining multiple molecular markers of excellent traits into one variety; (iv) realizing rapid accumulation of omics genotypes and phenomics [37].

Author details

Huixia Jia* and Haiping Wang*
Institute of Vegetables and Flowers, Chinese Academy of Agricultural Sciences, Beijing, China

*Address all correspondence to: jiahuixia@caas.cn and wanghaiping@caas.cn

References

[1] Sebastian P, Schaefer H, Telford IRH, Renner SS. Cucumber (*Cucumis sativus*) and melon (*C. melo*) have numerous wild relatives in Asia and Australia, and the sister species of melon is from Australia. Proceedings of the National Academy of Sciences of the United States of America. 2010;**107**(32): 14269-14273.

[2] Lv J, Qi JJ, Shi QX, Shen D, Zhang SP, Shao GJ, et al. Genetic diversity and population structure of cucumber (*Cucumis sativus* L.). Plos One. 2012;**7**(10):e46919.

[3] Qi JJ, Liu X, Shen D, Miao H, Xie BY, Li XX, et al. A genomic variation map provides insights into the genetic basis of cucumber domestication and diversity. Nature Genetics. 2013;**45**(12):1510-1515.

[4] Vora JD. Biochemical, Anti-microbial and organoleptic studies of cucumber (*Cucumis Sativus*). International Journal of Science & Research. 2014;**3**(3):662-664.

[5] Mukherjee PK, Nema NK, Maity N, Sarkar BK. Phytochemical and therapeutic potential of cucumber. Fitoterapia. 2013;**84**:227-236.

[6] Muruganantham N, Solomon S, Senthamilselvi MM. Anti-cancer activity of *Cucumis sativus* (cucumber) flowers against human liver cancer. International Journal of Clinical Pharmacology Research. 2016;**8**:39-41.

[7] Huang SW, Li RQ, Zhang ZH, Li L, Gu XF, Fan W, et al. The genome of the cucumber, *Cucumis sativus* L. Nature Genetics. 2009;**41**(12):1275-1281.

[8] Jamir MSharma A. A Sustainable production and marketing of cucumber crop in the hilly zone of Nagaland. TECHNOFAME-A Journal of Multidisciplinary Advance Research. 2014;**3**(1):61-66.

[9] Oboh G, Ademiluyi AO, Ogunsuyi OB, Oyeleye SI, Dada AF, Boligon AA. Cabbage and cucumber extracts exhibited anticholinesterase, antimonoamine oxidase and antioxidant properties. Journal of Food Biochemistry. 2017;**41**(3):e12358.

[10] Ibitoye OB, Uwazie JN, Ajiboye T. Bioactivity-guided isolation of kaempferol as the antidiabetic principle from *Cucumis sativus* L. fruits. Journal of Food Biochemistry. 2018;**42**(3):e12479.

[11] Fiume MM, Bergfeld WF, Belsito DV, Hill RA, Klaassen CD, Liebler DC, et al. Safety assessment of *Cucumis Sativus* (cucumber)-derived ingredients as used in cosmetics. International Journal of Toxicology. 2014;**33**:47-64.

[12] Moore JF, Duvivier R, Johanningsmeier SD. Formation of γ-aminobutyric acid (GABA) during the natural lactic acid fermentation of cucumber. Journal of Food Composition and Analysis. 2021;**96**:103711.

[13] Fleming HP. Developments in cucumber fermentation. Journal of Chemical Technology & Biotechnology Biotechnology. 2010;**34**(4):241-252.

[14] Girei A, Rabirou K, Ogezi E, Nnodu ON. Analysis of structure and performance of pumpkin (*CucurbitaPepo* L.) marketing in Nasarawa State. Direct Research Journal of Agricultural and Food Science. 2018;**6**(7):166-172.

[15] Futrell G, Others A. Introduction to agricultural marketing and Prices. Agricultural Production. 1983;**65**(4):114.

[16] Li YK, Xue XZ, Guo WZ, Wang LC, Duan MJ, Chen H, et al. Soil moisture and nitrate-nitrogen dynamics and economic yield in the greenhouse cultivation of tomato and cucumber under negative pressure irrigation in the North China Plain. Scientific Reports. 2019;**9**:4439.

[17] Zhang LD, Gao LH, Zhang LX, Wang SZ, Sui XL, Zhang ZX. Alternate furrow irrigation and nitrogen level effects on migration of water and nitrate-nitrogen in soil and root growth of cucumber in solar-greenhouse. Scientia Horticulturae. 2012;**138**:43-49.

[18] Fan ZB, Lin S, Zhang XM, Jiang ZM, Yang KC, Jian DD, et al. Conventional flooding irrigation causes an overuse of nitrogen fertilizer and low nitrogen use efficiency in intensively used solar greenhouse vegetable production. Agricultural Water Management. 2014;**144**:11-19.

[19] Qiu RJ, Kang SZ, Li FS, Du TS, Ling T, Wang F, et al. Energy partitioning and evapotranspiration of hot pepper grown in greenhouse with furrow and drip irrigation methods. Scientia Horticulturae. 2011;**129**(4):790-797.

[20] Zhang HM, Jin HJ, Ding XT, Yu JZ, Hao T. Efects of application of organic and inorganic fertilizers on the growth, yield and quality of cucumber in greenhouse. Journal of Plant Nutrition and Fertilizer. 2014;**20**(1):247-253.

[21] Böhme MH, Pinker I, Junge H. Effect of organic biostimulators with *Bacillus amyloliquefaciens ssp. plantarum* (Former *Bacillus subtilis*) as the main agent in vegetable cultivation. Bacilli and Agrobiotechnology. 2016;345-366.

[22] Dhillon MK, Singh R, Naresh JS, Sharma HC. The melon fruit fly, *Bactrocera cucurbitae*: a review of its biology and management. Journal of Insect Science. 2005;**5**(1):40.

[23] Abrahamian PE, Abou-Jawdah Y. Whitefly-transmitted criniviruses of cucurbits: current status and future prospects. Virusdisease. 2014;**25**(1):26-38.

[24] Gerling D, Alomar O, Arnó J. Biological control of *Bemisia tabaci* using predators and parasitoids. Crop Protection. 2001;**20**(9):779-799.

[25] Mahmood T, Tariq MS, Khokar KM, Hidayatullah, Hussain SI. Comparative effect of different plant extracts and insecticide application as dust to control the attack of red pumpkin beetle on cucumber. Pakistan Journal of Agricultural Research. 2010;**23**:196-199.

[26] Cradock KR, Graca JVD, Laing MD. Viruses infecting cucurbits in KwaZulu-Natal, South Africa. Revista Mexicana de Fitopatología. 2001;**19**(2):251-252.

[27] Ma JC, Du KM, Zheng FX, Zhang LX, Gong ZH, Sun ZF. A recognition method for cucumber diseases using leaf symptom images based on deep convolutional neural network. Computers and Electronics in Agriculture. 2018;**154**:18-24.

[28] Al-Sadi AM, Al-Said FA, Al-Quraini SM, kaabi SA, Al-Mazroui SS, Al-Mahmooli IHS, et al. Occurrence, characterization and management of fruit rot of immature cucumbers under greenhouse conditions in Oman. Phytopathologia Mediterranea. 2011;**50**(3):421-429.

[29] Mukhtar T, Muhammad ZK, Hussainc MA. Response of selected cucumber cultivars to *Meloidogyne incognita*. Crop Protection. 2013;**44**:13-17.

[30] Abu-Reidah IM, Arráez-Román D, Quirantes-Piné R, Arroyo SF, Carretero AS, Fernández-Gutiérrez A.

HPLC–ESI-Q-TOF-MS for a comprehensive characterization of bioactive phenolic compounds in cucumber whole fruit extract. Food Research International. 2012;**46**(1):108-117.

[31] Murcia MA, Jiménez AM, Martínez-Tomé M. Vegetables antioxidant losses during industrial processing and refrigerated storage. Food Research International. 2009;**42**(8):1046-1052.

[32] Peukert M, Weise S, Röder MS, Matthies IE. Development of SNP markers for genes of the phenyl propanoid pathway and their association to kernel and malting traits in barley. BMC Genetics. 2013;**14**:97.

[33] Feng SJ, Zhang JP, Mu ZH, Wang YJ, Wen CL, Wu T, et al. Recent progress on the molecular breeding of *Cucumis sativus* L. in China. Theoretical and Applied Genetics. 2020;**133**:1777-1790.

[34] Wang YH, Bo KL, Gu XF, Pan JS, Li YH, Chen JF, et al. Molecularly tagged genes and quantitative trait loci in cucumber with recommendations for QTL nomenclature. Horticulture Research. 2020;**7**:3.

[35] Wang YH, Jiang B, Dymerski R, Xu XW, Weng YQ. Quantitative trait loci for horticulturally important traits defining the Sikkim cucumber, *Cucumis sativus* var. *sikkimensis*. Theoretical and Applied Genetics. 2020;**134**(1):1-19.

[36] Zhang ZH, Map LY, Chen HM, Bu FJ, Li GC, Sun JJ, et al. Genome-wide mapping of structural variations reveals a copy number variant that determines reproductive morphology in cucumber. Plant Cell. 2015;**27**(6):1595-1604.

[37] Feng SJ, Zhang JP, Mu ZH, Wang YJ, Wen CL, Wu T, et al. Recent progress on the molecular breeding of *Cucumis sativus* L. in China. Theoretical and Applied Genetics. 2020;**133**(7): 1777-1790.

Section 2

Economic Values and Pharmacological Potentials

Chapter 2

Antidiabetic Principle in *Cucumis sativus* L.

Ibitoye Oluwayemisi Beatrice, Ajiboye Taofeek Olakunle, Abdulazeez Azeemat Titilola, Oyegoke Rukayat Abiodun, Muritala Hamdalat Folake and Abubakar Fatimah Aluko

Abstract

Diabetes is one of the leading cause of death globally. One of the strategies towards managing diabetes is the antidiabetic drugs which has recorded a huge success but accompanied with different degrees of side effect, hence, the use of natural plants products is encouraged. Several reports of antidiabetic medicinal plants have flooded literature but few has led to identification of active ingredient in such. *Cucumis sativus* is one of such plants reported to have antidiabetic property but there is little or no data on the active agent. This chapter therefore provides report on the active principle and mechanism of action underlying the antidiabetic activity of *C. sativus.*

Keywords: diabetes, *Cucumis sativus*, flavonoids, kaempferol

1. Introduction

Diabetes is a disorder where the body cells cannot use glucose effectively due to low insulin (Type 1 diabetes) or insulin insensitivity (Type 2 diabetes), therefore the blood glucose level increases [1]. It is characterized by a fasting blood glucose level higher than 126 mg/dL. It is one of the top 10 causes of death globally. About 463 million adults are living with diabetes; by 2045 this will rise to 700 million and Diabetes caused 4.2 million deaths in 2019 [2].

In 2017, total estimated cost of diagnosed diabetes in the U.S. was $327 billion [3].

Some complications of diabetes are oxidative stress, dyslipidaemia, endoplasmic reticulum (ER) stress [4, 5], retinopathy [6], neuropathy [7], nephropathy [8], cardiovascular complications [9], and ulcerations [10].

The management of diabetes has involved many approaches in order to enhance the availability of insulin, boost insulin sensitivity and reduce alpha glucosidase activity [11].

From research, people with excess weight can greatly manage diabetes by engaging in moderate and considerate weight loss plan, also exercise can help control blood sugar levels, reduce glycated hemoglobin and reduce insulin resistance.

Antidiabetic drugs are pharmacological substances that are employed in treating hyperglycemia when life style modifications do not bring desired effects [12]. They are categorized into different classes and they work either to enhance synthesis of insulin or reduce blood glucose level using different strategies.

These antidiabetic drugs are very effective in treating hyperglycemia, but despite this success, there has been increased side effects accompanying their use. Therefore, there is increased search for antidiabetic agents from medicinal plants with little or no adverse effects. Experimental reports have validated the presence of antidiabetic substances in medicinal plants [13–15]. One of such plants reported to have antidiabetic property is *Cucumis sativus*. Saidu *et al*.'s study reported the hypoglycemic property of methanolic fruit pulp extract of *Cucumis sativus* [16].

This chapter focuses on the antidiabetic principle identified in *Cucumis sativus* L.

2. Flavonoids

Flavonoids are a class of plants secondary metabolites made up of polyphenolic structures that contribute to the color and fragrance of fruits and flowers, therefore, they constitute a significant part of the human diet [17]. As a large class, flavonoids are subdivided into groups based on the structure of their carbon rings and these include flavanols, flavones, chalcones, flavonones, flavanonols and isoflavones. They are abundantly distributed in vegetables, fruits and some beverages. Flavonoids possess a wide range of health-promoting properties like the antioxidant effect, anti-carcinogenic, anti-inflammatory and antidiabetic capabilities. They display these properties by modulating the functions of some cellular enzymes as well as inhibition of different enzymes like lipo-oxygenase, cyclo-oxygenase, phosphoinositide 3-kinase and xanthine oxidase [18, 19]. Therefore, they are indispensable components in various pharmaceutical, cosmetics and medicinal applications. One of the flavonoids that possesses antidiabetic property is kaempferol.

3. *Cucumis sativus* L.

Cucumis sativus L. also known as Cucumber is a creeping plant in the family Cucurbitaceae. It is a fruit native to India and widely cultivated around the world. It is consumed fresh in salads, fermented (pickles) and as cooked vegetable [20].

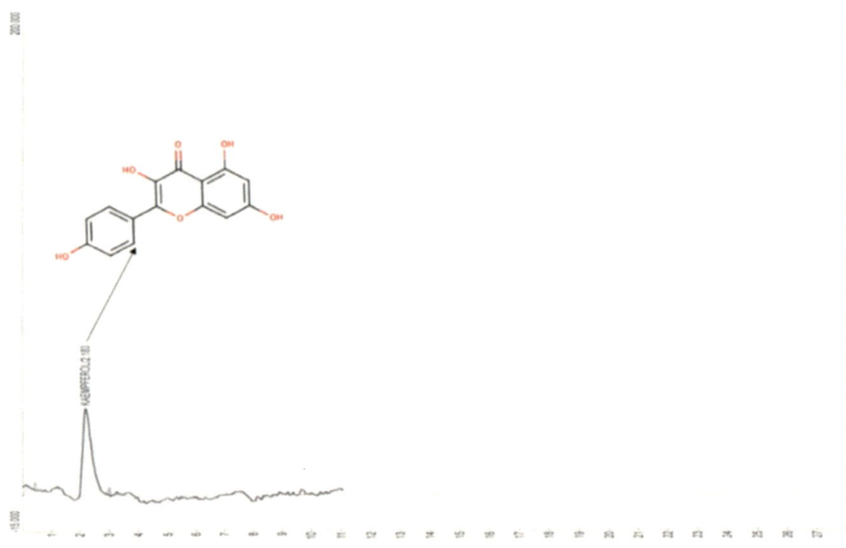

Figure 1.
High pressure liquid chromatographic profiling of the antidiabetic principle derived from Cucumis sativus fruit juice.

Cucumis sativus L. as a fruit, in addition to its nutritional value, has been reported to have some biological activities as anti-aging [21], antioxidant [22], and antidiabetic [23]. These properties have been linked to the presence of some phytochemical substances detected in *Cucumis sativus* L. like cucurbitacins [24], ascorbic acid [25], cucumerin, apigenin [26], lutein [27], quercetin 3-O-glucoside and kaempferol 3-O-glucoside [28].

Recent studies validated the presence of antidiabetic agents in *Cucumis sativus L.*

Ibitoye *et al* [29] identified the antidiabetic agent in *Cucumis sativus* L. as a flavonoid called kaempferol using HPLC (**Figure 1**).

4. Kaempferol

Kaempferol (3,5,7-trihydroxy-2-(4-hydroxyphenyl)-4H-1-benzopyran-4-one also known as kaempferol-3, **Figure 2**) is a yellow crystalline flavonoid having a molecular weight of 286.23 with a melting point of 276–278 °C. It is soluble in hot ethanol and slightly soluble in water.

It has been isolated from different parts of different plants. Yang *et al*. separated kaempferol and its derivatives from the methanolic crude extract of *Neocheiropteris palmatopedata* by repeated column chromatography, using a Sephadex LH-20 column [30]. Orhan et al. reported the bioactivity-guided fractionation of *Calluna vulgaris* and isolated kaempferol galactoside using successive column chromatography techniques [31]. Ibitoye et al. also reported the bioactivity guided isolation of kaempferol from *Cucumis sativus* L. [26].

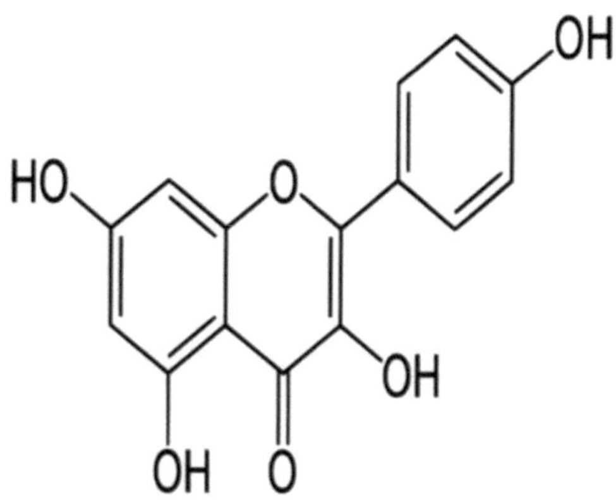

Figure 2.
Structure of Kaempferol.

5. Biosynthesis of kaempferol

Kaempferol and its derivatives are synthesized in plants by different types of enzymes. Kaempferol is synthesized by condensation of 4-coumaroyl-CoA with tripropionyl-CoA to produce naringenin chalcone, this reaction is catalyzed by chalcone synthase [32]. Naringenin chalcone is then converted into a flavanone called naringenin, which is thereafter hydroxylated by flavanone 3-dioxygenase to produce dihydrokaempferol [33]. Finally, the introduction of a double bond at the C2-C3 position of dihydrokaempferol produces kaempferol.

There is no much data on the pharmacokinetics of Kaempferol, however, flavonoids are extensively metabolized by the colonic microflora [34, 35].

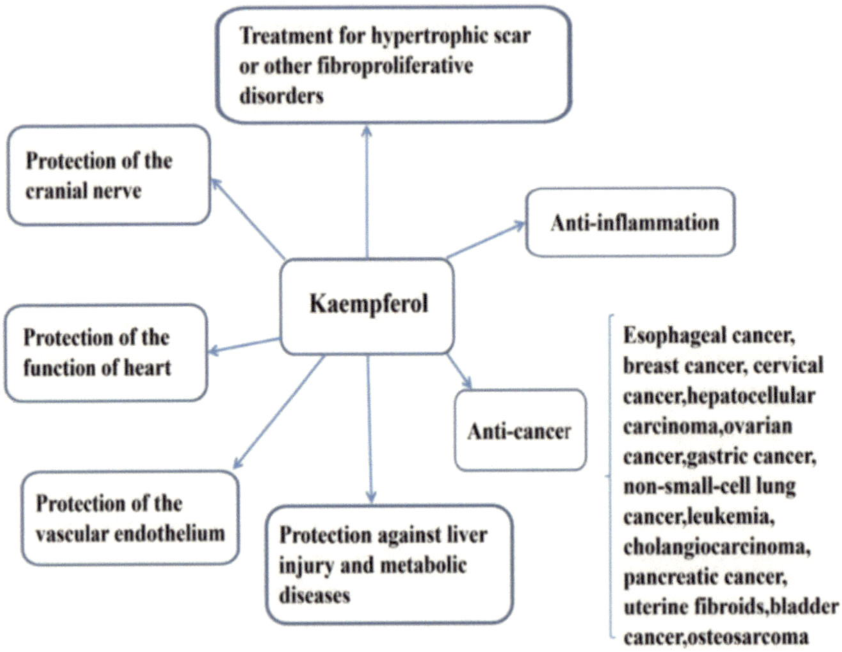

Figure 3.
Biological roles of Kaempferol.

Intestinal permeability study of kaempferol shows it undergoes significant biotransformation, with only a small fraction of the unchanged kaempferol able to cross the intestinal barrier [36].

It has been isolated from tea as well as common vegetables and fruits like beans, broccoli, cabbage, grapes, strawberries, tomatoes, apples and grapefruit [37].

Kaempferol has anti-inflammatory and anti-cancer properties, protects the liver and prevent metabolic diseases (**Figure 3**). The most well-known of its properties are its anti-inflammatory effects by decreasing lipopolysaccharide (LPS)-induced tumor necrosis factor-α (TNF-α) and interleukin-1 (IL-1) expression and also by increasing the number of activated macrophages [38]. Kaempferol is a dietary flavonoids that occur in fruits, vegetables, beverages, chocolates, herbs and plants [39] and reported to possess anti-diabetic property.

6. Mechanism of antidiabetic action of kaempferol

Kaempferol has been reported to lower blood glucose [40], inhibit α -amylase and α -glucosidase [41]. This section addresses the mechanism of action of Kaempferol under the following headings.

7. Inhibition of α-amylase and α-glucosidase enzymes.

α-Amylase and α-glucosidase are carbohydrate hydrolyzing enzymes located in the digestive tract. α- amylase in the duodenum initiates digestion and catalyzes the hydrolysis of α-1, 4 glycosidic linkages in starch resulting into sugars such as maltose, maltotriose and branched oligosaccharides. Then, α-glucosidase present in the brush border of the intestinal epithelium (enterocytes) is responsible for the final step of carbohydrates digestion, prior to their absorption. This enzyme cleaves terminal non-reducing 1, 4 linkages and converts the disaccharides and oligosaccharides into glucose, which is then transported by sodium/glucose co-transporter 1 (SGLT1) from the intestinal lumen to the cytosol of enterocytes. In turn, glucose transporter 2 (GLUT2), found in the basolateral membrane of enterocytes, transports glucose from cytosol to blood via facilitated diffusion.

One of the approaches to managing diabetes is to delay the absorption of glucose by the inhibition of carbohydrate hydrolyzing enzymes in the digestive tract of humans [42, 43].

Controlling the activity of these enzymes slows glucose production in the postprandial stage and this could be a therapeutic approach for people with diabetes. Hence, the search for inhibitors from medicinal plants is a great development [44].

Ibitoye et al. identified that kaempferol from *Cucumis sativus L.* lowers blood glucose and inhibited the activity of α -amylase and α glucosidase at IC_{50} of 51.24 and 29.37 μg/mL respectively [29]. This inhibition means reduction in blood glucose in the postprandial stage of alloxan-induced diabetic rats when given 165 mg/kg body weight of kaempferol from *Cucumis sativus* fruits. This evidently supports that kaempferol lowers blood glucose and inhibit α-amylase and α -glucosidase.

It may be possible that the glucose lowering activity of *C. sativus* fruits is through inhibition of α -amylase and α -glucosidase through kaempferol.

8. Maintaining glucose homeostasis

Diabetes features dysregulated glucose metabolism characterized by increased hepatic glucose production and decreased glucose oxidation. This eventually leads to deterioration in glucose control. Alkhalidy *et al* reported that kaempferol ameliorate hyperglycemia and enhance glucose tolerance in insulin deficient mice [45]. Diabetic mice displayed significantly higher pyruvate carboxylase activity. Kaempferol treatment suppressed the elevated pyruvate carboxylase activity and glucose-6 phosphatase activity in the liver suggesting that kaempferol may improve glycemic control in diabetes in part through suppressing gluconeogenesis in the liver via the regulation of pyruvate carboxylase, the first and critical step in gluconeogenesis [45]. It could therefore be a strategy for maintaining glucose homeostasis by targeting the glucose production and metabolic pathways.

9. Modulation of antioxidant profile

Generation of reactive oxygen species and free radicals contributes to the pathogenesis of diabetes [46]. This increased ROS production overruns the cellular antioxidant defense system leading to oxidative stress and damage [47]. Some diabetes research confirm this phenomenon in different diabetes model [48, 49]. Catalase, superoxide dismutase and glutathione are reduced significantly in diabetes [50]. Kaempferol reversed the alterations on oxidative stress markers in alloxan-induced diabetic rats [29].

10. Reversal of lipid profile alterations

One of the complications in diabetes is dyslipidemia, where the lipid profile is disturbed. It is usually presented with elevated levels of total cholesterol TC, triacylglycerol TAG, and low-density lipoprotein cholesterol LDLc and a reduction of high density lipoprotein cholesterol HDLc [51]. These alterations could predispose to developing atherosclerosis and cardiovascular diseases. Reversal of these alterations in alloxan-diabetic rats suggests its anti-dyslipidemic capability [29]. Alkhalidy et al. observed that untreated diabetic mice had lower total cholesterol, HDL-cholesterol, and LDL-cholesterol levels when compared to non-diabetic mice [45]. Kaempferol treatment reversed these changes to the levels similar to those seen in non-diabetic mice.

11. Maintenance of glycoprotein content

Glycoproteins are carbohydrate-containing proteins found on the cell membrane. They play important roles in membrane transport, cell differentiation and recognition, adhesion of macromolecules to cell surface and also in the secretion and absorption of macromolecules [52]. Impaired metabolism of glycoproteins contributes to the pathogenesis of diabetes [53]. Studies have reported that alterations in concentrations of various glycoproteins contribute to human diabetes [54]. Elevated levels of glycoproteins in diabetic condition could be a consequence of impaired carbohydrate metabolism [55]. Chandramohan *et al*. reported that kaempferol reversed elevated level of hexoses, hexosamines, fucose and sialic acid (glycoprotein componets) in streptozotocin-induced diabetic rats which may be due to the activation of glucose transport mechanism and also alters insulin binding receptor specificity [56].

Author details

Ibitoye Oluwayemisi Beatrice[1*], Ajiboye Taofeek Olakunle[2], Abdulazeez Azeemat Titilola[3], Oyegoke Rukayat Abiodun[4], Muritala Hamdalat Folake[4] and Abubakar Fatimah Aluko[4]

1 Antioxidants, Redox Biology and Toxicology Research Laboratory, Department of Biological Sciences, Al-Hikmah University, Ilorin, Nigeria

2 Antioxidants, Redox Biology and Toxicology Research Group, Department of Medical Biochemistry, College of Health Sciences, Nile University of Nigeria, Abuja, Nigeria

3 Department of Biological Sciences, Al-Hikmah University, Ilorin, Ilorin, Nigeria

4 Department of Biochemistry, Faculty of Life Sciences, University of Ilorin, Ilorin, Nigeria

*Address all correspondence to: inestimableyemi@gmail.com

References

[1] Zimmet, P., Alberti, K. G., & Shaw, J. (2001). Global and societal implications of the diabetes epidemic. Nature, 414, 782-787.

[2] The International Diabetes Federation Diabetes Atlas Ninth edition 2019. Diabetes facts & figures

[3] American Diabetes Association, The cost of Diabetes https://www.diabetes.org/resources/statistics/cost-diabetes Updated 02/2020. Accessed January 12, 2021.

[4] Ozcan, U., Cao, Q., Yilmaz, E., Lee, A.-H., Iwakoshi, N. N., Ozdelen, E., & Hotamisligil, G. S. (2004). Endoplasmic reticulum stress links obesity, insulin action, and type 2 diabetes. Science, 306, 457-461

[5] Poitout, V., & Robertson, R. P. (2008). Glucolipotoxicity: Fuel excess and cell dysfunction. Endocrine Reviews, 29(3), 351-366.

[6] Hove, M. N., Kristensen, J. K., Lauritzen, T., & Bek, T. (2004). The prevalence of retinopathy in an unselected population of type 2 diabetes patients from Arhus County, Denmark. Acta Ophthalmologica Scandinavica, 82, 443-448.

[7] Moran, A., Palmas, W., Field, L., Bhattarai, J., Schwartz, J. E., Weinstock, R. S., & Shea, S. (2004). Cardiovascular autonomic neuropathy is associated with microalbuminuria in older patients with type 2 diabetes. Diabetes Care, 27, 972-977.

[8] Huang, C., Kim, Y., Caramori, M. L., Fish, A. J., Rich, S. S., Miller, M. E., & Mauer, M. (2002). Cellular basis of diabetic nephropathy: II. The transforming growth factor-beta system and diabetic nephropathy lesions in type 1 diabetes. Diabetes, 51, 3577-3581.

[9] Saely, C. H., Aczel, S., Marte, T., Langer, P., & Drexel, H. (2004). Cardiovascular complications in Type 2 diabetes mellitus depend on the coronary angiographic state rather than on the diabetic state. Diabetologia, 47, 145-146.

[10] Wallace, C., Reiber, G. E., LeMaster, J., Smith, D. G., Sullivan, K., Hayes, S., & Vath, C. (2002). Incidence of falls, risk factors for falls, and fallrelated fractures in individuals with diabetes and a prior foot ulcer. Diabetes Care, 25, 1983-1986.

[11] Ajiboye, T. O., Uwazie, J., Haliru, F., Ibitoye, O., & Sunmonu, T. O. (2017). Mechanisms of action of columbamine, jatrorrhizine and magnoflorine as antidiabetic agents. In J. N. Govil (Ed.), Recent progress in medicinal plant (Vol. 45, pp. 221-232). USA: Studium Press LLC.

[12] Antidiabetic drugs. https://www.amboss.com/us/knowledge/antidiabetic_drugs Updated: October 5, 2020. Accessed: November 22, 2020.

[13] Marles, R. J., & Farnsworth, N. R. (1995). Antidiabetic plants and their active constituents. Phytomedicine: International Journal of Phytotherapy and Phytopharmacology, 2: 137-189.

[14] Sharma, N., & Garg, V. (2009). Antidiabetic and antioxidant potential of ethanolic extract of *Butea monosperma* leaves in alloxan-induced diabetic mice. Indian Journal of Biochemistry & Biophysics, 46: 99-105.

[15] Sun, H., Fang, W., Wang, W., & Hu, C. (2006). Structure-activity relationships of oleanane- and ursane-type triterpenoids. Botanical Studies, 47: 339-368.

[16] Saidu, A. N., Oibiokpa, F. I. & Olukotun, I. O. (2014). Phytochemical

screening and hypoglycemic effect of methanolic fruit pulp extract of *Cucumis sativus* in alloxan induced diabetic rats. Journal of Medicinal Plants Research, 8(39): 1173-1178.

[17] Wojdyo A, Oszmianski J, Czemerys R. (2007). Antioxidant activity and phenolic compounds in 32 selected herbs. Food Chemistry, 105:940-949.

[18] Panche, A., Diwan, A.D. & Chandra, S.R. (2016). Flavonoids: An overview. Journal of Nutrition Science, 5, 5.

[19] Metodiewa, D., Kochman, A. & Karolczak, S. (1997). Evidence for antiradical and antioxidant properties of four biologically active N, N-Diethylaminoethyl ethers of flavaone oximes: A comparison with natural polyphenolic flavonoid rutin action. IUBMB Life, 41, 1067-1075.

[20] Sotiroudis, G., Melliou, E., Sotiroudis, T. G., & Chinou, I. (2010). Chemical analysis, antioxidant and antimicrobial activity of three Greek cucumber (*Cucumis sativus*) cultivars. Journal of Food Biochemistry, 34: 61-78.

[21] Mukherjee, P. K., Maity, N., Nema, N. K., & Sarkar, B. K. (2011). Bioactive compounds from natural resources against skin aging. Phytomedicine, 19(1): 64-73.

[22] Nema, N. K., Maity, N., Sarkar, B., & Mukherjee, P. K. (2011). *Cucumis sativus* fruit-potential antioxidant, anti-hyaluronidase, and antielastase agent. Archives of Dermatological Research, 303: 247-252.

[23] Roman-Ramos, R., Flores-Saenz, J. L., & Alarcon-Aguilar, F. J. (1995). Anti-hyperglycemic effect of some edible plants. Journal of Ethnopharmacology, 48: 25-32.

[24] Enslin, P. R., Joubert, F. J., & Rehm, S. (1956). Bitter principles of the Cucurbitaceae. III.—Elaterase, an active enzyme for the hydrolysis of bitter principle glycosides. Journal of the Science of Food and Agriculture, 7: 646-655.

[25] Chu, Y.-F., Sun, J., Wu, X., & Liu, R. H. (2002). Antioxidant and anti-proliferative activities of common vegetables. Journal of Agricultural and Food Chemistry, 50: 6910-6916.

[26] McNally, D. J., Wurms, K. V., Labbe, C., Quideau, S., & Belanger, R. R. (2003). Complex C-glycosyl flavonoid phytoalexins from *Cucumis sativus*. Journal of Natural Products, 66: 1280-1283.

[27] Kai, H., Baba, M., & Okuyama, T. (2007). Two new megastigmanes from the leaves of *Cucumis sativus*. Chemical & Pharmaceutical Bulletin, 55: 133-136.

[28] Krauze-Baranowska, M., & Cisowski, W. (2001). Flavonoids from some species of the genus Cucumis, Biochemical Systematics and Ecology, 29(3), 321-324.

[29] Ibitoye, O. B., Uwazie, J. N., & Ajiboye, T. O. (2017). Bioactivity-guided isolation of kaempferol as the antidiabetic principle from *Cucumis sativus* L. fruits. *Journal of Food Biochemistry*, 12479.

[30] Yang, J.H., Kondratyuk, T.P., Marler, L.E., Qiu, X., Choi, Y., Cao, H., Yu, R., Sturdy, M., Pegan, S. & Liu, Y. (2010). Isolation and evaluation of kaempferol glycosides from the fern *Neocheiropteris palmatopedata*. Phytochemistry, 71, 641-647.

[31] Orhan, I., Küpeli, E., Terzioglu, S. & Yesilada, E. (2007). Bioassay-guided isolation of kaempferol-3-O-β-d-galactoside with anti inflammatory and antinociceptive activity from the aerial

part of *Calluna vulgaris* L. Journal of Ethnopharmacology, 114, 32-37.

[32] Devi, K. P., Malar, D. S., Nabavi, S. F., Sureda, A., Xiao, J., Nabavi, S. M. & Daglia, M. (2015). Kaempferol and inflammation: From chemistry to medicine. Pharmacology Research, 99:1-10.

[33] Tsao, R. (2010). Chemistry and biochemistry of dietary polyphenols. Nutrients, 2:1231-1246.

[34] Hollman, P.C.H. (2004). Absorption, bioavailability, and metabolism of flavonoids, Pharm. Biol. 42: 74-83.

[35] Van Duynhoven, J., Vaughan, E.E., Jacobs, D.M., Kemperman, R.A., Van Velzen, E.J.J., Gross, G., Roger, L.C., Possemiers, S., Smilde, A.K., Doré, J., Westerhuis, J.A. & Van de Wiele, T. (2011) . Metabolic fate of polyphenols in the human superorganism, Proc. Natl. Acad. Sci. U. S. A. 108: 4531-4538.

[36] Moradi-Afrapoli, F. Oufir, M., Walter, F.R., Deli, M.A., Smiesko, M., Zabela, V., Butterweck, .V. & Hamburger, M. (2016). Validation of UHPLC-MS/MS methods for the determination of kaempferol and its metabolite 4-hydroxyphenyl acetic acid, and application to in vitro blood-brain barrier and intestinal drug permeability studies, J. Pharm. Biomed. Anal. 128 264-274.

[37] Calderón-Montaño, J. M., Burgos-Morón, E., Pérez-Guerrero, C. & López-Lázaro M. (2011). A review on the dietary flavonoid kaempferol. Mini Review on Medicinal Chemistry, 11: 298-344.

[38] Ramachandran, V. and Baojun, X. (2015). Antidiabetic properties of dietary flavonoids: A cellular mechanism review, Nutr Metabol, 12: 60.

[39] Lin, M. K., Yu, Y. L., Chen, K. C., Chang, W. T., Lee, M. S., Yang, M. J., Cheng, H. C., Liu, C. H., Chen, D.C. & Chu, C. L. (2011). Kaempferol from *Semen cuscutae* attenuates the immune function of dendritic cells. Immunobiology, 216:1103-1109.

[40] Alkhalidy, H., Moore, W., Zhang, Y., McMillan, R., Wang, A., Ali, M. & Liu, D. (2015). Small molecule kaempferol promotes insulin sensitivity and preserved pancreatic b-cell mass in middle-aged obese diabetic mice. *Journal of Diabetes Research*, 532984.

[41] Peng, X., Zhang, G., Liao, Y. & Gong, D. (2016). Inhibitory kinetics and mechanism of kaempferol on α -glucosidase. Food Chemistry, 190207-190215.

[42] Hara, Y. & Honda, M. (1990). The Inhibition of α-Amylase by Tea Polyphenols. Agricultural and Biological Chemistry, 54, (8):1939-1945.

[43] Deshpande, M. C., Venkateswarlu, V., Babu, R. K. & Trivedi, R. K. (2009). Design and evaluation of oral bioadhesive controlled release formulations of miglitol, intended for prolonged inhibition of intestinal α-glucosidases and enhancement of plasma glucagon like peptide-1 levels. International Journal of Pharmaceutics, 380, (1-2): 16-24.

[44] Yin, Z., Zhang, W., Feng, F., Zhang, Y. & Kang, W. (2014). α-Glucosidase inhibitors isolated from medicinal plants. Food Science and Human Wellness, 33-4, 136-174.

[45] Alkhalidy, H., Moore, W., Wang, Y., Luo, J., McMillan, R. P., Zhen, W., Zhou, K. & Liu, D. (2018). The Flavonoid Kaempferol Ameliorates Streptozotocin-Induced Diabetes by Suppressing Hepatic Glucose Production. Molecules, 23, 2338.

[46] Rains, J. L., & Jain, S. K. (2011). Oxidative stress, insulin signaling,

and diabetes. Free Radical Biology & Medicine, 50, 567-575.

[47] Sies, H. (1991). Oxidative stress: From basic research to clinical application. The American Journal of Medicine, 91, 31S–38S.

[48] Oloyede, H. O. B., Ajiboye, T. O., Abdussalam, A. F., &. Adeleye, A. O. (2014). *Blighia sapida* leaves halt elevated blood glucose, dyslipidemia and oxidative stress in alloxan-induced diabetic rats. Journal of Ethnopharmacology, 157, 309-319.

[49] Oloyede, H. O. B., Bello, T. O., Ajiboye, T. O., & Salawu, M. O. (2015). Antidiabetic and antidyslipidemic activities of aqueous leaf extract of Dioscoreophyllum cumminsii (Stapf) Diels in alloxan-induced diabetic rats. Journal of Ethnopharmacology, 166, 313-322.

[50] Ajiboye, T. O., Akinpelu, S. A., Muritala, H. F., Ogunbode, S. M., Adeleye, A. O., Oladiji, A. T., & Oloyede, O. B. (2014). Trichosanthes cucumerina fruit extenuates dyslipidemia, protein oxidation, lipid peroxidation and DNA fragmentation in the liver of high-fat diet-fed rats. Journal of Food Biochemistry, 38, 480-490.

[51] Mittal, N., Keaur, J. & Mahmood, A. (1996). Changes in tubular membrane glycosylation in diabetic insulin and thyroxin treated rat kidneys. Indian Journal of Experimental Biology, 34: 782-785.

[52] Knecht KT, Bradford BU, Mason RP & Thurman RG. (1990). In vivo formation of free radicals metabolite of ethanol. Molecular Pharmacology, 38: 26-30.

[53] Sharma, C., Dalferes, F. R., Radhakrishnamurthy, B., De-Paolo, C. J. & Berenson, G. S. (1987). Hepatic glycoprotein synthesis in streptozotocin diabetic rats. Biochem Int, 1987; 36: 15-19.

[54] Chandramohan, G., Al-Numair, K. S., Alsaif, M. A. & Veeramani, C. (2015). Antidiabetic effect of kaempferol a flavonoid compound, on streptozotocin-induced diabetic rats with special reference to glycoprotein components. Progress in Nutrition, Vol. 17, N. 1: 50-57.

[55] Peppa, M., Stavroulakis, P. & Raptis, S. A. (2009). Advanced glycoxidation products and impaired diabetic wound healing. *Wound Repair Regen*eration, 17; 461-472.

[56] Marshall, S., Bacote, V. & Traxinger, R. R. (1991). Discovery of metabolic pathway mediating glucose induced desensitization of glucose transport system: role of hexosamine biosynthesis in the induction of insulin resistance. J Biol Chem 1991; 266: 4706-4712.

Chapter 3

Cucumber Pickles and Fermentations

Sarmad Ghazi Al-Shawi and Sadiq Jaafir Aziz Alneamah

Abstract

Cucumber sometimes used in sodium chloride solution as a substrate in lactic acid bacteria fermentation. The good fermentation always depending on many overlapped physical, chemical and microbial factors related with suspension of a strong and porous vegetables in a fluid. Keeping the cucumber integrity is very critical issue, and this may affect on the fermentation of liquid ingredients. This chapter tries to focus on the current efforts that conducting to control on the factors that affecting on cucumber fermentation. Modern and advanced technologies of recent studies are included within this chapter just like reducing the brining sodium chloride concentration, fresh cucumber gas exchange to develop their brining properties by using lactic acid bacteria cultures, developing an anaerobic cucumber fermentation tank system; preventing of cucumber gaseous spoilage by pouring of CO_2 from fermentation brines.

Keywords: lactic acid bacteria, cucumbers, fermentation, sodium chloride, brine

1. Introduction

1.1 Historical perspective

Different fermented foods could be categorized according to fermentation products just like organic acids which consisting of acetic acid and lactic acid (dairy and vegetables); and peptides and amino acids resulted from protein (fish and other fermentations); CO_2 (bread); and alcohol (wine and beer) [1, 2]. Food fermentation is one of an early the most precise innovations created and developed by people.

In Asia, coastal foragers during the age of primitive pottery (8000 to 3000 b.c.) were thought to have fermented vegetables before developing of crop-based agriculture [3]. It is possible that dairy fermentations in Middle East came after cattle domestication, alcohol was the first discovered fermented product from fruit fermentation. Many advanced fermentation procedures to produce alcohol by using the cereals were created nearly 4000 b.c., just like producing wine from rice in Asia and beer in Egypt [1]. In Asia, many composed references regarding fermentation innovation were found in historic poems Shijing Chinese book (1100 to 600 b.c.), that celebrates "the thousand wines of Yao," in referring to a kingdom in China from 2300 b.c. Cucumber thought were first fermented nearly 2000b.c. in Middle East. Old composed records came from the remains of papers of a play (The Taxiarchs) by Eupolis a writer from Greece (429–412 b.c.), also in Christian Bible, pickles were repeated many times. The fermented cabbage and kimchi on the Korean style, is expected to have established in the primitive pottery age from the wizened vegetables ordinary fermentation stored in seawater [1].

Sauerkraut on European style is thought was established in China, while the technique might be transferred to Europe at the invasion time of Mongol to central Europe in the 13th century. Nowadays, the vegetable fermentation industry is conducting on an enormous scale. Companies in United States that working on cucumber pickles fermentations may have 1000 fermentation tanks of forty-thousand-liter capacity at one site.

Throughout the ages, it was believed that cucumber pickle as the fairly fermented cucumber to which spices, vinegar, salt, and sometimes sugar has been added. While the preservation was not required by using the heat. Recently, fresh packed pickles, manufactured by adding of spices, salt, and vinegar to the fresh cucumbers under pasteurized preservation, are representing a huge portion of pickle industry.

Industrial treatments tentatively preserve around 40% of crop through the fermentation in NaCI brines that contain fermentable carbohydrates which converting into acetic acids, ethanol, lactic acid, CO_2, and other compounds by naturally existence lactic acid bacteria and yeast. This procedure uses to expand the using equipment packing line and workers to throughout the year operation in manufacturing of the final product.

Traditionally, fiberglass, wood, and polyethylene tanks are used for the fermentation that might require 10–21 days (period of storage in the same tanks is generally less than 1 year) and sometimes longer. Tanks are put outdoors to give the opportunity for sunlight ultraviolet irradiation to hit the surface of the brine and subsequently inhibiting yeasts and molds growing, and other microorganisms on the surface of the brine.

During the fermentation of cucumber pickles, brine storage and processing operations are liable to the reactions of oxidation which affect adversely on the quality properties. In spite of pickles are flooded in brine during fermentation and bulk-storage, while the containers are opening, which encourage the exposure to air and sunlight.

Additionally, pickle tanks' brines are usually spread with air in order to mix the components and to release CO_2, and at the time of transferring to processing operations, pickles are removed from brine and subsequently exposes to light and oxygen. Also, the brines and pickles content of traces prooxidant metals just like copper, zinc and iron which act together with oxygen and light to be in charge of pigments oxidation and developing undesirable flavors sometimes, and this may lead to considerable economic loss of the market value.

1.2 Cucumber fermentations

Cucumber (*Cucumis sativus*) fermentation in United States is conducted in 30,000–40,000 liter, fiberglass tanks with open top and placed outdoors to allowing the surface to exposure to sunlight. Sunlight UV radiation is dependent to suppression the surface aerobic yeasts that have the ability to utilize lactic acid that resulted from fermentation. Cucumbers are submerged totally with salt brine and kept under the brine surface with wooden headboards. Fermentation is usually conducted in 6% NaCl. Calcium chloride typically added the cover brine in order to keep the fragile texture, and firm of the fermented cucumber throughout fermentation and storage [4]. The fermentations of cucumber usually subject a homolactic acid fermentation, that is not resulting CO_2 from sugars. Although CO_2 could be produced via cucumbers respiration and via malate decarboxylation over the beginning of fermentation [4]. Some of lactic acid bacteria have an analytical malolactic enzyme that converting malate to lactate and CO_2. The reaction of malolactic enzyme takes place intracellularly resulting in proton absorption, subsequently increasing the internal pH of the cell. Although it is a recommendable reaction in winemaking (applied to removing the acidity of wines), the fermentation of malolactic in cucumbers may lead to formation of "bloaters," or undesirable pockets of internal gas, resulting in decreasing the yield of the production [5]. In order to prevent the formation of bloater, the fermentation of cucumber is clean with air to get rid the surplus CO_2 from the tank [6]. In order to restrict the growing of aerobic microorganisms in air-purged cucumber fermentations, especially molds and yeasts, acetic acid (0.16%) or potassium sorbate (~0.04%) could be used as aids in processing [7].

Air purging may be stopped each day several hours to control aerobic micro-organisms' excessive growth. Usually, cucumber is fermented with *Lactobacillus. plantarum* and other LAB and may store for year in fermentation tanks in degrees under than 0 °C while NaCl concentration commonly increase to 10–15% during the storage to reduce freezing damage and keeping the required fermented cucumber texture. Cucumber should be washed before selling in order to remove the excess salt and then using different packages (jars, pouches, plastic pails) with suitable covers in packaging. The covers usually contain spices, acetic acid, and lactic acid residues. Pasteurization sometimes is used for fermented pickles while heat treatment is not used for big containers. Excessive growth of microorganisms is eliminated by low pH, organic acids, and absence of fermentable sugars. Cucumber fermentations depend on the growing of LAB that existence naturally on cucumbers surface. Although, some starter cultures are added to cucumber fermentation to get a consistent product, adding *Lactobacillus plantarum* does not decarboxylate malic acid (subsequently does not form bloaters) [8], and this approach has been created, developed, and tasted to identify culture growing capability in cucumber fermentations [9]. A procedure for starter culture preparation that suitable for the requirements of kosher is applicable to producers [10]. The brined cucumbers' primary pH is nearly 6.5. Recycled brine could be used in commercial fermentations, or adding acetic acid to brine solutions. This acid addition may help in removing the excess CO_2 and encouraging LAB growth, so the commercial fermentations' primary pH could vary basically. Some of the metabolites could have an inhibitory effect on the other bacteria just like peroxides, bacteriocins, and peptides [11]. There might be 1.5% lactic acid, pH (3.1–3.5), few or no sugar at the end of fermentation. In such an environment that is acidic, anaerobic, high salty, and lacks sugar, there are a low number of microorganisms that have the ability to grow and survive to preserve cucumbers. Sometimes during storage, fermented cucumbers expose secondary undesired fermentation which is identified by pH increasing, lactic acid vanishing, propionic and butyric acid formation. Deterioration of fermented cucumber happening at the spring season beginning when increasing the surrounding temperature. Increasing propionic and butyric acid concentrations lead to smelly spoilage [12]. The microbial environment of this spoilage presently is not closely defined but may attribute to the growth of bacteria that form spores such as clostridia when increasing the pH above 4.6. The salt concentration of the fermented cucumbers is about (6% or more) and this is very high for consumption directly by humans. Therefore, the salt concentration is reduced to around 2% by water washing directly before packing and distribution. This treatment lead to high salt concentrations of the waste stream in addition to a high BOD resulting from the organic ingredients that are existed in the brine and that spread out of cucumbers over the process of desalting. Hence, cucumber brine of the desalting process is commonly recycled and might be utilized another fermentation [13]. The brines fermentation could be treated in order to expel the softening enzymes (mostly polygalacturonases) before the recycling [14], which acts on degrading cucumber cell's pectic substances and softening the fruits.

1.3 Critical factors for fermenting cucumbers

Fermentation is influenced by variables due to cucumbers, environmental conditions under which they are kept during fermentation, and microorganisms that are naturally present or intentionally added. Since it is so important to maintain the structural integrity of cucumbers, both physical and chemical factors are involved. The interactions between these factors lead to an extremely interesting and complex fermentation process [15]. A lot of research on the fermentation of cucumbers and other fruits and vegetables has been done. However, there is an incomplete understanding of the interactions between the microbiological, chemical, and physical factors involved.

Before the cucumber fermentation industry can take full advantage of the biotechnology revolution that looms for many fermentation industries, more understanding of these interactions is needed [16].

1.4 Microbial changes in spoilage

The production of CO_2 in the cover brine of fermenting vegetables by heterofermentative LAB and fermentative species of yeasts has been linked with gas pockets formation inside the cucumber, which called formation of bloater (**Figure 1**). Homofermentative LAB capable of decarboxylating malic acid, as example *L. plantarum*, might cause bloating by producing a sufficient CO_2 when combined with the CO_2 formed from the respiring vegetable tissues [8, 18]. Prevention of bloater formation was effective in fermented cucumber brines by using nitrogen or air [6, 19]. Air purging has to be carefully controlled as it may result in fruit softening due to mold growth [20, 21] reduced brine acidity due to yeast growth and off-colors and flavors. The addition of potassium sorbate to fermentation brines, including the application of spray to brine surfaces, is widely used to minimize the growth of yeast and the development of CO_2.

Oxidative yeasts may cause malodorous spoilage of fermented cucumbers to develop. The lactic acid generated during fermentation can be consumed by these microorganisms, with a subsequent increase in pH that facilitates the development of spoilage microorganisms [22, 23]. In cucumbers, lactic acid produced during primary fermentation can be catabolized by yeasts of the genera Pichia and Issatchenkia, causing an increase in pH.

Pectinolytic enzymes derived from plant material or microbes can cause the softening of brined vegetables (**Figure 2**).

Mold growth accompanying film-forming yeast growth on the brine surface can cause softening of cucumbers. In the absence of sunlight and the presence of minimal amounts of oxygen, heavy scum yeast and/or mold growth is generally the result of neglecting brine material during extended storage [25]. In order to maintain anaerobic conditions and to limit the growth of surface yeasts and molds, Pickled cucumber tanks are usually held indoors, with a seated plastic cover weighted down with water or brine. Mold polygalacturonases associated with cucumber flowers can also result in the softening of brined cucumbers [26]. By draining and rebrining the tank with calcium chloride, this problem can be reduced. 36 hours after the initial brining procedure. However, this solution is not about salt disposal. Recycled brines are instead treated to inactivate the softening enzymes, if necessary [14]. The addition of calcium chloride may slow down the rate of fermenting cucumbers' enzymatic softening. This should not, however, be relied upon

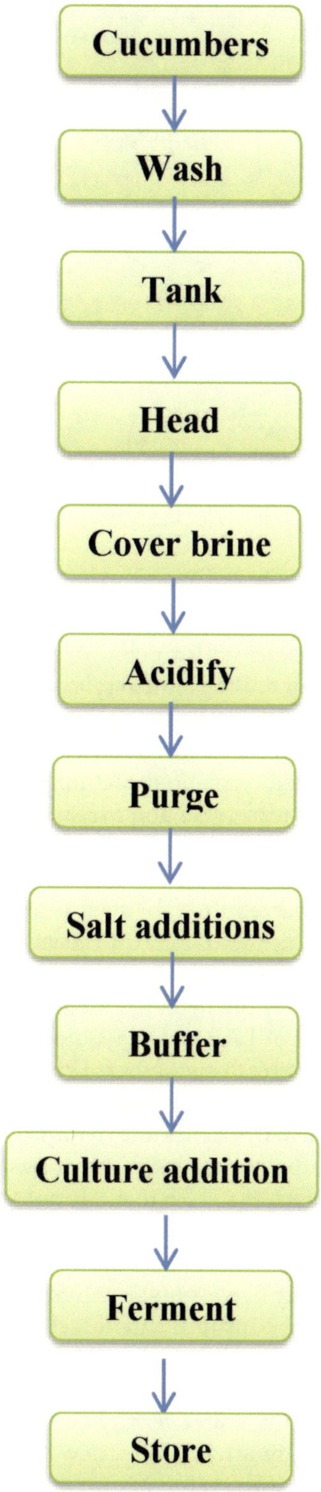

Figure 1.
Steps brine fermentation of cucumber [17].

to eliminate problems with enzymatic softening. Care must be taken to minimize the contamination of flowers and plant debris by cucumbers, especially small fruits, which may be a source of contamination by pectinolytic molds. Due to the reduced

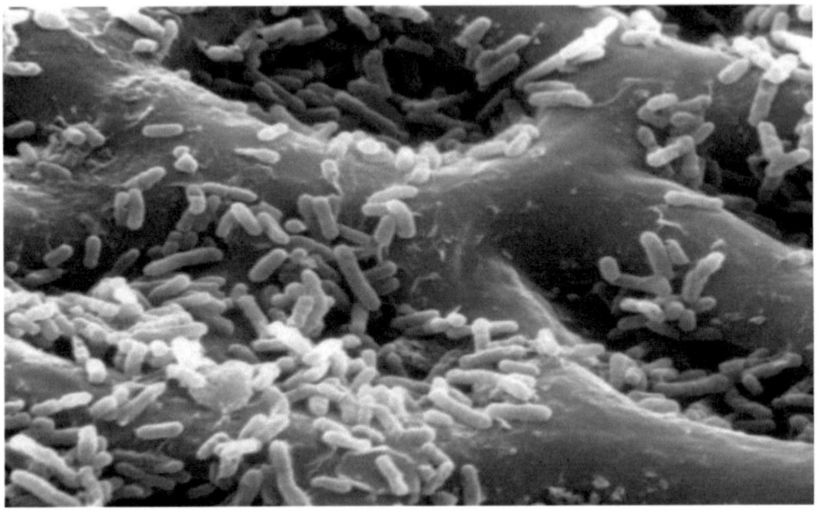

Figure 2.
Lactobacillus plantarum cells colonizing the cucumber tissue [24].

amount of brine surface in contact with air compared to the total volume, softening is not a very serious problem in bulk Spanish-style cucumber fermentation. Yeasts and/or molds on the plastic drums used during the conditioning operations (sizing, grading, pitting, stuffing, etc.) can, however, cause softening [22]. Desalting is used to prepare non-pasteurized fermented cucumbers, followed by the addition of cover liquor, often containing acetic acid and preservatives. Sugar is added to sweet pickles at concentrations of up to 40 percent. The main spoilage organisms in such products are osmotolerant yeasts, and a preservation prediction chart, based on the concentration of acid and sugar required for shelf stability, has been developed. On the surface of the liquid, aerobic molds and film yeasts may grow, mainly as a result of defective jar closure. Spoilage microorganisms in sweet pickles include yeasts [27] and lactobacilli, particularly the heterofermentative *Lactobacillus fructivorans*. In order to prevent the growth of LAB and yeast, non-fermented pickle products in which acetic acid is added to fresh cucumbers (known as fresh-pack pickles) are pasteurized. Recommended procedures include 165 °F (74 °C) for 15 minutes, as described by [28]. Spoilage usually occurs due to improper processing (insufficient heat to pasteurize) and/or improper acidification of pasteurized pickle products, so that a balanced brine product of pH 3.8 to 4.0 is not achieved. Molds and film yeasts are factors in cases of poor jar closure, where oxygen is introduced into the container, as with sweet pickles.

This can lead to a potentially dangerous situation triggered by an increase in pH as the spoilage microorganisms consume organic acids. Germination of *Clostridium botulinum* spores can occur if the pH rises above 4.6. Non-acidified refrigerated products are sold commercially under a variety of names, including half-sour dills, real kosher dills, new kosher dills, sour overnight dills, garlic pickles, new half-sour pickles, new half-sour pickles, new half-sours, new home-style pickles, etc. [29]. These cucumbers may be kept at room temperature in barrels for a few days or longer and then refrigerated at 2–5 ° C to allow fermentation to occur. Microbial growth, enzymatic activity, and the curing process continue at a slow rate under cooling conditions [29]. The gaseous spoilage of the product is caused primarily by the previously mentioned microbial groups that form gas. Due to the much lower concentrations of salt added to these product types, softening issues in refrigerated-fermented products may develop. To such products, fresh, whole garlic cloves and other spices are normally added. It is possible that these spices contain softening enzymes. Whether the half-sour products are manufactured in bulk or in the retail jar, for more than a few weeks, the very nature of the product makes it difficult to maintain good quality. The barreled product achieves the Good Manufacturing Practices (GMP)-recommended brine pH of 4.6 or lowers for acidified foods typically before or shortly after refrigeration, and then slowly begins to produce acid. For a product made in a retail jar, this recommended condition for brine-product pH cannot be ensured because there is no uniform process adopted by the packers in which the product is initially acidified or intentionally incubated for the development of natural fermentation with lactic acid.

The refrigerated fresh-pack (non-fermented) products contain 2–3 percent NaCl and sometimes sodium benzoate or other preservatives and are acidified with vinegar at a balanced pH of around 3.7 [29]. The cucumbers are not heated, like the half-sour pickles, either before or after packing. The products will maintain an acceptable quality for several months if properly acidified, refrigerated, and preserved. However, recipes containing no vinegar or other acid in the initial cover liquor should be considered with caution. Quality assurance of cucumber products begins with the removal of the cucumber's outer leaves and woody core. In addition to its undesirable texture, the existing sucrose in the core could be utilized by *Lactobacillus mesenteroides* resulting in formation of dextran which lead to a stringy and slimy texture. Cucumbers marketed under refrigerated conditions are preserved by the addition of sodium benzoate and metabisulfite [30]. Chemical changes that can result in discoloration (browning) and the formation of objectionable flavors influence the shelf life of such products. The growth of naturally occurring yeasts in cucumbers may result from uneven salting during cucumber preparation and may induce pink coloring and vegetable softening. Spanish-style olives were formerly preserved in cover solutions containing relatively high salt concentrations through fermentation. However, it has been demonstrated that an appropriate combination of low pH (3.5), combined acidity (0.025) mill equivalents (mEq)/L) and moderate proportions of acid (>20.4%) and salt (>25.0%) is also able to preserve well-cured cucumbers [31]. Incompletely cured cucumbers or those with characteristics outside the ranges necessary for complete stabilization without heat treatment have been gradually used to allow pasteurization to be commercialized [22]. In some cases, particularly when pasteurization is not recommended (plastic bags, seasoned olives, etc.), producers used authorized preservatives such as potassium sorbate or sodium benzoate [31].

1.5 Fermentations biochemistry

Usually, fermentation is defined as an anaerobic process. Within the cucumber fermentation process, LAB and yeast convert glucose and fructose into lactic acid, ethanol, acetic acid, and CO_2. The homofermentative LAB main pathway is breaking down of one six-carbon sugar (glucose) to produce two molecules of three-carbon lactic acid. More complex metabolism is used by Heterofermentative organisms. At the beginning, glucose is converted into CO_2 five-carbon sugar phosphate, and furthermore degraded into lactic acid and a two-carbon compound, acetic acid or ethanol [32]. We shall concentrate here on vegetable fermentation biochemical features that link to quality of the product. So far, many researches are paying more attention in vegetables fermentation and storage, especially cucumbers, with reduced salt. Vegetable fermentations' chloride waste can be extremely reduced in case of reducing the required salt for fermentation and storage in order to exclude the desalting step before the conversion to final products. Many research studied the relationship between concentration and type of the salt [33]. Replacing of NaCl with various cations and anions on fermentation of sugar in cucumber juice. The most interesting thing, fructose was the most preferred fermentable sugar to *Lactobacillus plantarum* more than glucose in most of experiments. Along with addition of different salts, the utilization of sugar was decreasing as anion or cation concentrations increasing. Lu et al. [33] and Zhou et al. [34] have identified various volatile ingredients in cucumbers that fermented with *Lactobacillus plantarum* (2% NaCl). About 37 volatile ingredients were determined, and as a result of fermentation, there was a little change in most of them. Inhibition of (E, Z)-2,6-nonadienal and 2-nonenal production was the most outstanding fermentation effect on cucumber volatiles [35]. Characterized trans- and cis-4-hexenoic acid as the strongest odors that specify the brine aroma properties of commercially fermented cucumbers in nearly 6% NaCl [36]. Illustrated that exposing the slurries (2% NaCl) of fermented cucumber to oxygen resulting in formation of nonenzymatic hexanal plus a series of trans unsaturated aldehydes with 5–8 carbon atoms that linked with oxidized odor intensity development the tissue of fermented cucumber. In the existence of light, about 100 μg/ml concentration of calcium disodium EDTA preserve nonfermented pickles against bleaching of pigments and lipid oxidation [14]. Although, when using this compound, there was a little reduction in pickles' firmness retention. Firmness retention in cucumbers fermentation and storage is a key quality issue. It is difficult to assure the firmness retention (in reduced salt fermented cucumbers) equal to what can be accomplished by fermenting and storage in 6% NaCl or more. Nevertheless, over many previous years there was a wide understanding for softening of cucumber tissue.

Fleming et al. [21] showed the importance of calcium in keeping fermented cucumbers' firmness. It was found that first-order kinetics is followed by the nonenzymatic softening of acidified, blanched cucumber tissue [37]. The mentioned kinetic manner made it reasonable to identify the activation of entropy and enthalpy of cucumbers' nonenzymatic softening, although that the chemical reactions in charge of softening were not known. At 1.5 M NaCl, both activation of entropy and enthalpy were high. Cucumber softening was inhibited by calcium because it reduced activation entropy too much into a limit that activation overall free energy was reduced [38]. This behavior of thermodynamic is resembled to that which occurs when changing conformation of polymers, just like in denaturation of protein. It is totally differed from the observed properties of pectin acid hydrolysis [39]. Figured out that pectin's acid hydrolysis rate was inefficient to be the reason for non-enzymatic softening the tissue of the cucumber [40]. Identified salt, temperature, and calcium concentrations combined effects on fermented cucumber tissue's softening rate. The softening kinetics of fermented cucumbers did not follow the first-order simple reaction. Just like the tissues of many other plants, cucumber possesses enzymes that have the ability to degrade the ingredients of plant cell walls, which may lead to changing in the texture.

In cucumbers, many activities of enzymes have been found such as exopolygalacturonase, pectinesterase, and endopolygalacturonase [41]. When fermenting or acidifying of cucumber, methyl groups are removed from pectin by pectinesterase [42]. Nevertheless, pectin's' enzymatic hydrolysis by polygalacturonases from cucumber has not been identified if it is a significant factor in fermented cucumbers' softening. Adding of fungal polygalacturonases into the tanks of fermentation, especially on the flowers attached to small cucumbers has been linked to the commercially importance of fermented cucumbers' enzymatic softening. Buescher and Burgin [43] developed a sensitive new method of diffusion plate to determine the activity of polygalacturonase in the brines of fermentation and found that alumino-silicate clay has the ability of adsorbing and removing the activity of polygalacturonase from the brines of fermentation that are recycled. Enzymes which could hydrolyze polysaccharides of the cucumbers cell wall have not studied widely comparing with the enzymes that degrade pectin [45]. Showed that the activity of endo-β-1,4- gluconase in cucumber is inhibited under pH of 4.8 while endoglucomannan-splitting enzyme retains its activity under pH of 4.0 but is inhibited within the fermentation. In fresh cucumbers, they characterized 6 enzymes which hydrolyze p-nitrophenylglycosides of β-d-glucose, β-d-galactose, α-d-galactose, β-d-xylose, α-d-mannose, and α-l-arabinose, which were inhibited throughout the fermentation. The enzymes that have the ability to hydrolyze the synthetic substrates are widespread in plants. Resemble enzymatic activities were found in olives, pears, and Semillon grapes.

Maruvada [44] and Takayanagi et al. [45] discovered the same p-nitrophenyl glycosidases detected by [44] in cucumbers. She reported undetectable levels in 2% NaCl brines throughout the first week of fermentation [46, 47]. Gathered calcium addition, fresh cucumbers' blanching relatively to enzyme inactivation, and a quick fermentation using a malolactic-negative *Lactobacillus plantarum* culture for cucumbers' fermentation and keeping a required texture in reduced (4%) sodium chloride concentration [48]. Found notable degradation products of glucosinolate in cucumbers fermented with *Lactobacillus sakei* compared to cucumbers manufactured with lactic acid bacteria starter cultures [49]. Reported that ascorbigen, a compound resulted from a degradation product reaction of indole glucosinolate (glucobrassicin) and ascorbic acid, is the cucumbers' dominant glucosinolate degradation product. Glucoraphinin existed in fresh cucumbers was converted over the fermentation into sulforphorane, however, sulforphorane was a relatively small glucosinolate degradation product in fermented cucumbers. There are many concerns about the biogenic amines' formation in cucumbers [50]. Reported that storing cucumbers up to 12 months led to the formation of tyramine. While very trace amounts of tryptamine, histamine, and spermine were determined. These findings were assured in a study on vegetable products which concluded that tyramine concentration was about 4.9 mg/100 g in canned cucumbers [51], and the same finding and the concentration reported by [50]. No health risk existed referring to these mentioned biogenic amine levels, with the possible exception that individuals taking medications possessing monoamine oxidase inhibitors.

1.6 Fermented cucumbers-related problems

Compared to the fermentation of liquids such as beer, wine, and milk, unique problems are involved in the fermentation of whole vegetables. Structural integrity has to be preserved in whole vegetables, which is not a factor with liquids [52]. Tissue softening is also a serious defect that can be caused by pectinolytic enzymes of either microbial (primarily fungal) source [53] or of the cucumber fruit itself. Off-flavors and off-colors may result from improper methods of fermentation and handling.

The cucumber pickle industry is faced with waste disposal, in addition to spoilage problems. These wastes consist of the salt used to prevent softening during fermentation and storage, and the organic wastes. Salt concentrations used greatly exceed the 2–3 percent desired in the final product [54].

Thus, after storing the brine, the excess salt must be leached from the cucumbers before they are processed into finished products. Disposal of this non-biodegradable waste salt is a source of serious environmental concern. As the salt is extracted during leaching, soluble cucumbers, including desirable nutrients and flavor compounds, are also removed. These desirable components are not only lost, they must be degraded before being discharged into waterways. Discharge of salt and organic materials into municipal disposal systems typically entails an extra expense for pickle companies, since municipalities must charge for recovering the cost of handling such waste [55] (**Figure 3**).

Figure 3.
Cucumber bloater defect caused by carbon dioxide microbiologically produced during fermentation by either yeasts or LAB [56].

1.7 Preventing flavor formation

Purge-and-trap analysis of cucumber slurries' volatile ingredients in 2 percent reduced-salt salt brine before and after cucumber fermentation. Volatile components' comparison before and after fermentation led to the derivation that the main influence of fermentation on volatile flavors was to prohibit the enzymatic production of E, Z-2,6-nonadienal and 2-nonenal enzymes in cucumbers [34]. These aldehydes are the major ingredients in charge of cucumbers' fresh flavor [57]. Although, after a few days of cucumber fermentation, when tearing the tissue of cucumber, the pH descends low enough to deactivate the enzymes that forming these compounds. In fresh cucumber slurries, just benzaldehyde, ethyl benzene, and o-xylene were not found within the volatile ingredients characterized in the fermented cucumbers. Recently, the absence of flavor influence of volatile aldehydes is the main effect of the fermentation on flavor [35]. In fermented pickled cucumber brines, a low influence of volatility flavor compound was characterized. Adding of saturated salt to brine samples and heating to 50 °C, SPME (solid-phase microextraction) fiber sampling followed by GC-olfactometry resulted in the identification of a component with an odor close to that of the fermentation brine. The component with a fermentation brine odor was characterized as *trans*-4-hexenoic acid. The existence of cis-4-hexenoic acid was also tentatively characterized. A solution containing 25 ppm trans-4-hexenoic acid, 10 ppm phenyl ethyl alcohol, 0.65 percent lactic acid, 0.05 percent acetic acid, and 8 percent sodium chloride in a reconstitute experiment had an odor very similar to that of fermented cucumber brine. Lactic acid, acetic acid, and sodium chloride concentrations are acceptable for commercial brines after completing the fermentation. Adding of phenyl ethyl alcohol resulted in in a few enhancements in the matching odor. For that, the key component in the simulated brine solution was trans-4-hexenoic acid. The source of trans-4-hexenoic acid in fermentation brines is, unfortunately, not recognized.

Author details

Sarmad Ghazi Al-Shawi[1]* and Sadiq Jaafir Aziz Alneamah[2]

1 Food Science Department, Agriculture College, Basrah University, Basrah, Iraq

2 Food Science Department, Agriculture College, Kufa University, Al-Najaf, Iraq

*Address all correspondence to: sarmadghazi@yahoo.com

References

[1] Lee, C. H. 2001. Fermentation Technology in Korea. Korea University Press, Seoul, South Korea.

[2] Steinkraus, K. H. 1993. Comparison of fermented foods of the East and West, p. 1-12. In C. H. Lee, K. H. Steinkraus, and P. J. A. Reilly (ed.), Fish Fermentation Technology. UNU Press, Tokyo, Japan.

[3] Lee, C. H. 2009. Food biotechnology, p. 85-114. In G. Campbell-Platt (ed.), Food Science and Technology. Wiley-Blackwell, West Sussex, United Kingdom.

[4] Fleming, H. P., K.-H. Kyung, and F. Breidt, Jr. 1995. Vegetable fermentations, p. 629-661. In H.-J. Rehm and G. Reed (ed.), Biotechnology. VCH Publishers, New York, NY.

[5] McFeeters, R. F., H. P. Fleming, and R. L. Thompson.1982. Malic acid as a source of carbon dioxide in cucumber fermentations. J. Food Sci. 47:1862-1865.

[6] Costilow, R. N., C. L. Bedford, D. Mingus, and D. Black. 1977. Purging of natural salt-stock pickle fermentations to reduce bloater damage. J. Food Sci. 42:234-240.

[7] Gates, K., and R. N. Costilow. 1981. Factors influencing softening of salt-stock pickles in air-purged fermentations. J. Food Sci. 46:274-277.

[8] Daeschel, M. A., R. F. McFeeters, H. P. Fleming, T. R. Klaenhammer, and R. B. Sanozky. 1984. Mutation and selection of Lactobacillus plantarum strains that do not produce carbon dioxide from malate. Appl. Environ. Microbiol. 47:419-420.

[9] Breidt, F., and H. P. Fleming. 1992. Competitive growth of genetically marked malolactic-deficient Lactobacillus plantarum in cucumber fermentations. Appl. Environ. Microbiol. 58:3845-3849.

[10] Perez-Diaz, I. M., and R. F. McFeeters. 2010. Preservation of acidified cucumbers with a natural preservative combination of fumaric acid and allyl isothiocyanate that target lactic acid bacteria and yeasts. J. Food Sci. 75: M204–M208.

[11] DeVuyst, L., and E. J. Vandamme. 1994. Antimicrobial potential of lactic acid bacteria, p. 91-142. In L. DeVuyst and E. J. Vandamme (ed.), Bacteriocins of Lactic acid Bacteria. Blackie Academic and Professional, London, United Kingdom.

[12] Fleming, H. P., M. A. Daeschel, R. F. McFeeters, and M. D. Pierson. 1989. Butyric acid spoilage of fermented cucumbers. J. Food Sci. 54:636-639.

[13] McFeeters, R. F., W. Coon, M. P. Palnitkar, M. Velting, and N. Fehringer. 1978. Reuse of Fermentation Brines in the Cucumber Pickling Industry, p. 1-115. EPA-600/2-78-207. U.S. Environmental Protection Agency, Washington, DC.

[14] Buescher, R., and C. Hamilton. 2002. Adsorption of polygalacturonase from recycled cucumber pickle brines by Pure-Flo B80 clay. J. Food Biochem. 26:153-156.

[15] Medina, E., M. Brenes, C. Romero, A. Garcia, and A. de Castro. 2007. Main antimicrobial compounds in table olives. *J. Agric. Food Chem.* 55:9817-9823.

[16] Tolonen, M., S. Rajaniemi, J.-M. Pihlava, T. Johansson, P. E. J. Saris, and E.-L. Ryhanen. 2004. Formation of nisin, plant-derived biomolecules and antimicrobial activity in starter culture fermentations of sauerkraut. *Food Microbiol.* 21:167-179.

[17] Fleming, H. P. 1982. Economic Microbiology. Fermented foods (Rose,A. H., Ed.), Academic press, New York.

[18] Fleming, H. P., W. M. Walter, Jr., and J. L. Etchells. 1973. Antimicrobial properties of oleuropein and products of its hydrolysis from green olives. Appl. Microbiol. 26:777-782.

[19] Fleming, H. P., J. L. Etchells, R. L. Thompson, and T. A. Bell. 1975. Purging of CO_2 from cucumber brines to reduce bloater damage. J. Food Sci. 40:1304-1310.

[20] Costilow, R. N., K. Gates, and M. L. Lacy. 1980. Molds in brined cucumbers: Cause of softening during air purging of fermentations. Appl. Environ. Microbiol. 40:417-422.

[21] Fleming, H. P., R. L. Thompson, T. A. Bell, and L. H. Hontz. 1978. Controlled fermentation of sliced cucumbers. J. Food Sci. 43:888-891.

[22] Garrido Ferna'ndez, A., M. J. Ferna'ndez Dı'ez, and R. M. Adams. 1997. Table olives: production and processing. Chapman & Hall, London, UK.

[23] Gonza'lez Cancho, F., L. Rejano Navarro, and J. M. R. Borbolla y Alcala. 1980. Formation of propionic acid during the conservation of table green olives. III. Responsible microorganisms. Grasas y Aceites 31:245-250.

[24] Reina, L. D., H. P. Fleming, and F. Breidt, Jr. 2002. Bacterial contamination of cucumber fruit through adhesion. J. Food Prot. 65:1881-1887.

[25] Etchells, J. L., and T. A. Bell. 1950. Film yeasts on commercial cucumber brines. Food Technol. 4:77-83.

[26] Bell, T. A., J. L. Etchells, and R. N. Costilow. 1958. Softening enzyme activity of cucumber flowers from northern production areas. Food Res. 23:198-204.

[27] Dakin, J. C., and P. M. Day. 1958. Yeasts causing spoilage in acetic acid preserves. J. Appl. Bacteriol. 21:94-96.

[28] Etchells, J. L., T. A. Bell, H. P. Fleming, R. E. Kelling, and R. L. Thompson. 1973. Suggested procedure for the controlled fermentation of commercially brined pickling cucumbers – The use of starter cultures and reduction of carbon dioxide accumulation. Pickle Packers Science. 3:4-14.

[29] Etchells, J. L., T. A. Bell, and W. R. Moore Jr. 1976. Refrigerated dill pickles--Questions and answers. Pickle Pak Sci. 5:1-20.

[30] Stamer, J. R., and B. O. Stoyla. 1978. Stability of sauerkraut packaged in plastic bags. J. Food Prot. 41:525-529.

[31] Arroyo-Lo'pez, F. N., J. Bautista-Gallego, K. A. Segovia- Bravo, P. Garcı'a-Garcı'a, M. C. Dura'n-Quintana, C. Romero, F. Rodrı'guez-Go'mez, and A. Garrido-Ferna'ndez. 2009. Instability profile of fresh packed 'seasoned' Manzanilla- Aloren˜ a table olives. LWT Food Sci. Technol. 42:1629-1639.

[32] Gottschalk, G. 1986. *Bacterial Metabolism*, 2nd ed., p. 208-220. Springer-Verlag, New York, NY. 45. Heredia, A., R. Guillen, A. Jimenez, and J. Fernandez- Bolanos. 1993. Activity of glycosidases during development and ripening of olive fruit. *Z. Lebensm. Unters. Forsch*. 196:147-151.

[33] Lu, Z., H. P. Fleming, and R. F. McFeeters. 2002. Effects of fruit size on fresh cucumber composition and the chemical and physical consequences of fermentation. *J. Food Sci*. 67:2934-2939.

[34] Zhou, A., and R. F. McFeeters. 1998. Volatile compounds in cucumbers fermented in low-salt conditions. *J. Agric. Food Chem*. 46:2117-2122.

[35] Marsili, R. T., and N. Miller. 2000. Determination of major aroma impact

compounds in fermented cucumbers by solid-phase microextraction-gas chromatography-mass spectrometry-olfactometry detection. *J. Chromatogr. Sci.* 38:307-314.

[36] Zhou, A., R. F. McFeeters, and H. P. Fleming. 2000. Development of oxidized odor fermented cucumber tissue exposed to oxygen. *J. Agric.* and volatile aldehydes in *Food Chem.* 48:193-197.

[37] McFeeters, R. F., and H. P. Fleming. 1989. Inhibition of cucumber tissue softening in acid brines by multivalent cations: inadequacy of the pectin "egg box" model to explain textural effects. *J. Agric. Food Chem.* 37:1053-1059.

[38] McFeeters, R. F., and H. P. Fleming. 1990. Effect of calcium ions on the thermodynamics of cucumber tissue softening. *J. Food Sci.* 55:446-449.

[39] Krall, S. M., and R. F. McFeeters. 1998. Pectin hydrolysis: effect of temperature, degree of methylation, pH, and calcium on hydrolysis rates. *J. Agric. Food Chem.* 46:1311-1315.

[40] McFeeters, R. F., M. Balbuena, M. Brenes, and H. P. Fleming. 1995. Softening rates of fermented cucumber tissue: effects of pH, calcium, and temperature. *J. Food Sci.* 60:786-788.

[41] McFeeters, R. F., T. A. Bell, and H. P. Fleming. 1980. An endo-polygalacturonase in cucumber fruit. *J. Food Biochem.* 4:1-16.

[42] McFeeters, R. F., and S. A. Armstrong. 1984. Measurement of pectin methylation in plant cell walls. *Anal. Biochem.* 139:212-217.

[43] Buescher, R. W., and C. Burgin. 1992. Diffusion plate assay for measurement of polygalacturonase activities in pickle brines. *J. Food Biochem.* 16:59-68.

[44] Maruvada, R. 2005. Evaluation of the importance of enzymatic and non-enzymatic softening in low salt cucumber fermentations. M.S. thesis. North Carolina State University, Raleigh.

[45] Takayanagi, T., T. Okuda, and K. Yokotsuka. 1997. Changes in glycosidase activity in grapes during development. *J. Inst. Enol. Viticult. Yamanashi Univ.* 32:1-4.

[46] Meurer, P., and K. Gierschner. 1992. Occurrence and effect of indigenous and eventual microbial enzymes in lactic acid fermented vegetables. *Acta Aliment.* 21:171-188.

[47] Fleming, H. P., E. G. Humphries, R. L. Thompson, and R. F. McFeeters. 2002. Bag-in-box technology: storage stability of process-ready, fermented cucumbers. *Pickle Pak Sci.* 8:14-18.

[48] Medina, E., C. Gori, M. Servili, A. de Castro, C. Romero, and M. Brenes. 2010. Main variables affecting the lactic acid fermentation of table olives. *Int. J. Food Sci. Technol.* 45:1291-1296.

[49] Ciska, E., and D. R. Pathak. 2004. Glucosinolate derivatives in stored fermented cabbage. *J. Agric. Food Chem.* 52:7938-7943.

[50] Kalacˇ, P., J. Spicka, M. Krizek, and T. Pelikanova. 2000. Changes in biogenic amine concentrations during sauerkraut storage. *Food Chem.* 69:309-314.

[51] Moret, S., D. Smela, T. Populin, and L. S. Conte. 2005. A survey on free biogenic amine content of fresh and preserved vegetables. *Food Chem.* 89:355-361.

[52] Fleming, H. P.; Pharr, D. M.1980 j. food Sci.,45, 1595-1600.

[53] Eschells, J. L.; Bell, T. A.; Monroe, R. J.; Masley, P. M.; Demain, A. L. 1958 Apple. Microbiol. 6, 427-440.

[54] Rybaczyk-Pathak, D. 2005. Joint association of high cabbage/ sauerkraut intake at 12-13 years of age and adulthood with reduced breast cancer risk in Polish migrant women: results from the U.S. Component of the Polish Women's Health Study (PWHS), abstr. 3697. *Abstr. Am. Assoc. Cancer Res. 4th Annu. Frontiers Cancer Prevention Res.*, Baltimore, MD, 2 November 2005.

[55] Tolonen, M., T., T. Marianne, V. Britta, P. Juha-Matti, K. Hannu, and R. Eeva-Liisa. 2002. Plant-derived biomolecules in fermented cabbage. *J. Agric. Food Chem.* 50:6798-6803.

[56] Fleming, H. P. 1979. Purging carbon dioxide from cucumber brines to prevent bloater damage--A review. Pickle Pak Sci. 6(1):8.

[57] Schieberle P, Ofner S, Grosch W. 1990. Evaluation of potent odorants in cucumbers and muskmelons by aroma extract dilution analysis. J Food Sci 55:193-5.

Chapter 4

Cosmetic, Culinary and Therapeutic Uses of Cucumber (*Cucumis sativus* L.)

Chidiebere Ugwu and Stephen Suru

Abstract

Cucumber (*Cucumis sativus* L.) is cultivated throughout the world as an important vegetable. This review offers an insight on the versatile use of cucumbers for cosmetic, culinary and therapeutic purposes. Epidemiological and nutritional studies have shown various benefits associated with the use of cucumber. As cosmetic, cucumber is popularly used for natural beautification and for skin treatments. As a vegetable, cucumber is the quintessential materials and indispensable for salads, soups and smoothie with diverse health benefits including weight loss, remedy for chronic constipation, anti-inflammatory, cardiovascular and cancerous diseases among others. The use including weight loss, cosmetic, culinary and therapeutic purposes.

Keywords: cucumber, culinary, cosmetics, therapeutic, uses, health, benefits

1. Introduction

Cucumber (*Cucumis sativus* L.) is cultivated throughout the world because of its ability to thrive in both temperate and tropical conditions. As a result, fresh cucumbers are available throughout the year [1]. Historically, cucumber is one of the oldest cultivated crops and believed to be a native of the Asia continent [2]; parts of China with temperate climates and parts of southern regions of India with tropical climates. At present, cucumber is the fourth most widely cultivated vegetable crop in the world (after tomatoes, onions and cabbage) and China is by far the world's largest producer of it [1].

C. sativus belongs to the gourd family of Cucurbitaceous, which also includes cantaloupes, squash, pumpkins, melon and water melon [3, 4]. *C. sativus*is a tender creeping vine and hairy leaves with 3 to 5 pointed lobes. In general, it bears roughly cylindrical fruits and may be as large as 60 cm long and 10 cm in diameter with dark-green skin, crispy moisture rich flesh, and small edible seeds concentrated at its core [5]. There are close to 100 varieties, but common ones include the English, garden, Persian, mini, and lemon. Cucumbers are mainly eaten in the unripe green form when they taste sweet, have crunchy texture, and unique flavor. Thus, they are best-harvested young, tender and just short of reaching maturity. On getting to full maturity, the cucumber skin becomes tougher and turns yellow-white with an accompanied bitter and sour [1, 5].

Relative to other vegetables that have been widely researched and reported on, cucumber seems to have received little interest probably because of seemingly lack of concentrated levels of well-known bioactive compounds presents in garlic, onion, tomatoes and the likes. Generally, cucumbers are consumed because of their refreshing quality, in part due to their very high water content and crunchy texture. Studies have shown that cucumbers contain lignans, vitamin K, cucurbitacins and their derivatives (triterpenoids), flavonoids (apigenin, luteolin, quercetin, and kaempferol), antioxidants (beta carotene and vitamin C) and B vitamins among other trace elements and minerals [6–8]. With a 95% level of water content saturated with naturally-occurring nutrients and trace elements meshed in high dietary fiber, cucumbers are beginning to gain attention in therapeutic, culinary and cosmetic uses.

2. Cosmetic uses of cucumber

Cucumber has various benefits associated with skin treatments and natural beautification. As a cosmetic, cucumber is excellent for rubbing over the skin to keep it soft and white [9]. Because of its nutritious and an extremely cooling property, cucumbers are used by women to bringing cooling relief to the eyes in summertime [10]. Cucumber slices offer many topical benefits to the eyes and surrounding tissues through their hydrating properties, which work to reduce dehydration and their high levels of vitamin K that help reduce cutaneous eruption such as puffiness (eye bags) and dark shadows [9]. Cucumbers contain lignans that help to soothe down irritation and inflammation associated with sun burns and insect bites [11].

Cucumber fruit extracts are often incorporated as a primary ingredient in many topical skin preparations. Such preparations have been used as a moisturizer and skin toner by inhibiting tyrosinase [12]. Additionally, such preparations have been used to treat wrinkles and cleanse the skin. The deep cleansing action of cucumbers emanates from its naturally occurring organic acids such as glycolic, lactic and salicylic acids [9].

Glycolic and lactic acids are alpha hydroxyl acids used as chemical exfoliants that promote the natural removal of dead cells and to keep the protective surface layers healthy by dissolving the glue-like substance in the epidermal layer. The glue-like substance causes a buildup of dead skin cell layer; leaving skin dehydrated, dull and coarse. Histologically, alpha hydroxy acids have been shown to increase the thickness of the epidermis as well as cause increased collagen density, improved elastic fiber quality, increased papillary dermal thickness and increased dermal acid mucopolysacharide translating into thicker, healthier skin with fewer rhytids [13–15]. Alpha hydroxyl acids have been effective in the treatment of many skin conditions such as ache, psoriasis, bumps, pustules, eczema, dry skin, age spots, seborrheic keratosis, precancerous growths, hyperkeratosis, actinic keratosis and also black heads and whiteheads [16].

Glycolic acid is the most active and beneficial of the alphahydroxyl acids in skin care, because of its ability to penetrate through the cell wall by virtue of its small molecular size [17]. Once inside the cell, it triggers new formation of collagen and turns on the synthesis of dermal glycosaminoglycans to plump up the cell and the ground substance in the skin to reduce wrinkles on the skin's surface [18]. Lactic acid improves the appearance of photodamage and surface pigmentation [19].

Unlike glycolic and lactic acids, salicylic acid is a beta hydroxyl acid that exhibits a keratolytic, antiseptic and fungicidal properties [20]. It can be used for the treatment of hyperkeratotic and scaling conditions such as dandruff, ichthyosis and psoriaisis [21]. The fungicidal properties of salicylic acid may partly explain the topical use of cucumber preparation in the treatment of fungal skin infections such as tinea [22].

Cucumber soap is used by many women, and a cucumber wash applied to the skin after exposure to keen winds is extremely beneficial. It is used in preparation of glycerin and cucumber cream. Cucumber has use in perfume production [9]. Overall, cucumber and its preparations have become part of daily beauty product into face packs, facials, juice and many other things which can affect your skin [1].

3. Culinary uses of cucumber

Cucumber is the quintessential fruit that can be added to a variety of dishes. Typically, they are indispensable for salads or used for pickling, soups and smoothie especially in warm season or summer. Cucumber wedges tossed in a garden salad (consumed with fried and barbequed foods), slices on a sandwich, or used as an appetizer for parties have become the mainstay of many of today's lunches [1].

There are basically three separate uses for cucumbers: fresh whole, fresh sliced, and pickled. Regardless of variety, fresh whole cucumbers are grown for consumer retail sales. On the other hand, fresh sliced cucumbers are typically garden variety and are grown for the foodservice sector, which requires uniform sized slices for packaged salads and restaurant chain salad bars.

Pickling cucumbers tend to be smaller and thicker. The best known variety is the bumpy-skinned gherkin. Not all pickled forms of cucumbers are fermented. Fermented pickled cucumbers are made by combining cucumbers with water, salt, and bacteria and giving the bacteria the right amount of time to convert various substances in the cucumbers into different bioactive compounds that can elicit health benefits especially in the gastrointestinal tract. However, the unfermented pickled cucumbers, though labeled as pickles and usually sold in grocery, are made by submerging cucumbers in a very acidic liquid (usually vinegar).

The consumption of fresh whole cucumbers has become a common trend among middle-class households in some African cultures. The fresh whole cucumbers are eaten alone as a snack or in conjunction with peanuts or peanut butter as an appetizer during folkloric/traditional display of hospitality.

In recent times, the cucumber diet have become increasingly popular and included among many sought-after choices because of its availability throughout the season, low calorific value and high dietary fiber as well as stress-free procedure associated with the conventional short-term weight reduction therapies [1]. In principle, there are no standard rules to this diet other than replacing most foods with cucumbers, along with a few protein-rich foods, such as eggs, chicken, lean meat, fish, cottage cheese and nuts. Since the diet lacks variety, it is considered extremely restrictive and unsustainable for a long-term with an attendant health risk.

4. Therapeutic uses of cucumber

Traditionally, cucumber has been used in folk medicine to treat diseases such as diarrhea, diabetes, and hypertension. Consisting mostly water in which numerous electrolytes and phytochemicals are saturated, the unique chemical profile of cucumbers is thought to elicit a number of possible health benefits Notwithstanding, that some of these claims are still undergoing investigation, there is proven evidence that the phytoconstituents in cucumber possess chemopreventive and anticancer properties, antioxidant and anti-inflammatory properties [23–26].

Cucumber has also been reported to modify plasma lipid and act as an analgesic [27]. Cucumbers are good sources of more than 73 different phenolic compounds known to elicit health benefits.

4.1 Hydration

On a daily basis, adequate hydration is crucial for healthy living and prevention of diseases especially the likes of constipation and kidney stones. Cucumber is a good source of superior hydration due to high content of water (95%) saturated with naturally-occurring electrolytes. Consumption of cucumbers offers the cells the much-needed hydration and vital nutrients required for optimal cellular functioning, repair and maintenance of membrane integrity [28, 29]. Thus cucumbers can help prevent dehydration during summer time or during and after exercises. Cucumber extract when pharmacologically refined has been reported to have the potential in the treatment of corneal acid burn through its hydrating properties [9].

4.2 Body weight management

Cucumbers have high content of dietary fiber and very low calorific value resulting from low carbohydrates and very low amount of protein and fat contents. Consumption of cucumbers can help to heighten satiety and naturally curb appetite, which make it easier to cut down on food intake. Cucumbers also improve digestion because of their high fiber content, and adequate digestion has been linked with easier weight loss. In view of the aforementioned, the cucumber diet has emerged and has been included among many sought-after choices in short-term weight loss therapeutic regimens [1]. However, it should be noted that cucumber diet are usually restrictive in variety and as such must be complemented with some protein-rich foods.

4.3 Bone health

Cucumber contains calcium, phosphorus and vitamin K. Vitamin K helps to improve calcium absorption. A sufficient intake of these elements has been associated with maintenance of healthy bones that are less likely to fracture especially among the elderly. Put together, these nutrients contribute to good bone health.

4.4 Management of blood glucose level

Cucumbers have a low score on glycemic index (GI), indicating that they provide important nutrients without or with minimal carbohydrates that can cause a spike in blood glucose level [30]. Besides, a recent report suggested that cucurbitans in cucumber can stimulate the release of insulin and regulate the metabolism of a key hormone in the processing of blood glucose and hepatic glycogen [31]. Put together, this may suggest that cucumbers may help in the control and prevention of diabetes, given credence to folkloric claim.

4.5 Potential anti-cancer health benefit

Cucumbers contain high levels of triterpen family of phytochemicals known as cucurbitacins. The consumption of cucumber avails us cucurbitacins A, B, C, D, and E, which may help prevent cancer by stopping cancer cells from proliferating and surviving. Cancer research studies have proved that the JAK–STAT and MAPK signaling pathways involved with cancer cell development and survival can be blocked through the effects of cucurbitacins [24, 25]. While there are currently no current anti-cancer therapies that utilize cucurbitacins, experimental research has yielded promising results awaiting confirmation in human studies [31, 32].

The results of cancer studies have shown that cucumber lignans such as lariciresinol, pinoresinol and secoisolariciresinol are converted by intestinal bacteria into enterolignans. These enterolignans including enterodiol and enterolactone have been reported to bind to estrogen receptors thereby eliciting both pro-estrogenic and anti-estrogenic effects [33]. Some preliminary results have shown that consumption of plant derived lignans including cucumbers can reduce estrogen-related cancer risk of the ovary, prostate, breast and uterus [34].

4.6 Cardiovascular health benefits of cucumber

There are several processes through which cucumber consumption may elicit cardiovascular benefits. Cucumbers are good sources of dietary fiber, particularly in their skins. Dietary fibers are known to significantly reduce the absorption of dietary cholesterol thus positively modifying the blood lipid profile with an attendant reduction in cholesterol buildup in the arteries. More so, cucumbers provide potassium and magnesium that may contribute to preventing high blood pressure [9, 35, 36]. With 73 different phenolic constituents, cucumber provides protection against oxidative insults for the blood vessels and their vulnerable contents such as low density lipoproteins among others [26].

The cucurbitacins in cucumber may also help prevent atherosclerosis. There have been reports on Cucurbitacin B and E in glycosidic form to exhibit inhibitory effect on lipid oxidation products like malondialdehyde (MDA) and 4-hydroxynonenal (4-HNE) [37, 38]. These positive reports bolster the therapeutic role of cucurbitacins in artherosclerosis, which involves modification of lipoproteins by involvement of- MDA and 4-HNE [39].

4.7 Antioxidant and anti-inflammatory benefits of cucumber

A good number of phytochemicals present in cucumber have been reported to exhibit antioxidant and/or anti-inflammatory activities via sparing effect on other antioxidants and/or regulation of antioxidant enzymes in metabolic pathways involved. Small-scale human studies have been conducted using some of these identified phytochemicals in cucumber and found to provide some health benefits [29]. The cyclooxygenase 2 (COX 2), a pro-inflammatory enzyme, has been shown to be inhibited by cucumber extract [34]. In addition, antioxidant enzymes like superoxide dismutase, catalase and glutathione peroxidase showed increased activities in studied participants that consumed cucumber powder. Some studies have shown that cucumber can reduce the generation of reactive oxygen species (ROS) and reactive carbonyl species (RCS), which could be of help in individuals with type 2 diabetes. The high content of phenolic compounds (flavonoids and terpenoids) present in cucumbers may be involved in this probable health benefit by significantly lowering the levels of ROS and RCS. Its high content of fisetin flavonoid gives it the potential anticancer, antioxidant and anti-inflammatory benefits [40]. Since cucumbers are rich in phytonutrients, there could be logical theoretical link between their consumption and amelioration of some chronic diseases.

4.8 Other health benefits of cucumber

The peel and seeds are the nutrient dense parts of cucumber and contain beta-carotene that is good for the eyes [41]. Cucumber provides an alkaline diet, and due to its triterpene content, works well in regulating diseases of the immune system [42]. Majority of alkaline fruits and vegetables like cucumber are also anti-inflammatory in nature [43] and thus neutralize the body's acid pH specifically in the kidneys [29].

5. Conclusions

The cosmetic and culinary uses of cucumber without any reported adverse effect are indicative of its versatility. Cucumbers are endowed with phytochemicals that have been reported to elicit positive health benefits ranging from hydration of body cells, body weight control and management of degenerative diseases. It is hoped that as more clinical researches are conducted on whole extracts and phytoconstituents, amazing health benefits shall be unfolded and folkloric claims shall be clarified for diverse but specific therapeutic uses.

Conflict of interest

The authors declare no conflict of interest.

Author details

Chidiebere Ugwu* and Stephen Suru
Department of Human Biochemistry, Faculty of Basic Medical Sciences, College of Health Sciences, Nnamdi Azikiwe University, Anambra State, Nigeria

*Address all correspondence to: ce.ugwu@unizik.edu.ng

References

[1] Maheshwari RK, Mohan L, Malhotra J, Updhuay B, Rani B. Invigorating efficacy of Cucumissativus for healthcare and radiance. Int. J. Chem. Pharmaceut. Sci., 2014; 2(3):737-744.

[2] Abulude OA, Adeleke KO. Comparative studies on nutritional composition of four melon seeds varieties. Pak J Nutr. 2010; 9:905-908.

[3] Bello MO, Owoeye G, Abdulhammed M, Yekeen TA. Characterization of Gourd fruit (cucubitaceae) for dietary values and anti-nutrition constituent. J Pharm BiolChem Sci. 2014; 6:7575-7585.

[4] Vivek KB, Ji-Eun K, Yong-Ha P, Sun CK. In-vivo pharmacological effectiveness of heat-treated cucumber (Cucumis sativus L.) juice against CCl4 - induced detoxification in a rat model. Indian Journal of Pharmaceutical Education and Research. 2017; 51(2):280-287.

[5] Doijode SD. Seed storage of horticultural crops. Edn. Haworth Press; 2001. 339p.

[6] Grubben GJH, Denton OA. Plant Resources of Tropical Africa 2 Vegetables, Netherlands: Leiden, Wageningen, Backhuys Publishers; 2004. 48-57p.

[7] Wang YH, Joobeur T, Dean RA, Staub, JE Cucurbits-genome mapping and molecular breeding in plants 5. Vegetables.2007; 375.

[8] Mukherjee PK, et al. Phytochemical and therapeutic potential of cucumber. Fitoterapia.2013; 84:227-236.

[9] Uzodike EB, Onuoha IN. The effect of cucumber (*Cucumbissativus*) extract on acid induced corneal burn in guinea pigs. JNOA.2009; 15:3-7.

[10] Mateljan G. The World's healthiest foods-essential guide for the healthiest way of eating. 1st ed..World's Healthiest Foods; 2006. 880p.

[11] Lopes L, Speretta F, Bentley M. Enhancement of skin penetration of vitamin K using monoolein-based liquid crystalline systems. European Journal of Pharmaceutical Sciences. 2007; 32(3):209-215.

[12] Hooda R. Antiwrinkle herbal drugs. Journal of Pharmacognosy and Phytochemistry. 2015; 4(4): 277-281.

[13] Moy LS, Howe R, Moy RL. Glycolic acid modulation of collagen production in human skin fibroblast cultures in vitro. Dermatol. Surg. 1996; 22 (5): 439-441.

[14] Kim SJ, Park JH, Kim DH, Won YH, Maiback HI. Increased in vivo collagen synthesis and in vitro cell proliferative effect of glycolic acid.Dermatol. Surg. 1998; 24 (10): 1054-1058.

[15] Bernstein EF, Lee J, Brown DB, Yu R, Vanscott E. Glycolic acid treatment increases type 1 collagen mRNA and hyaluronic acid content of human skin. Dermatol. Surg. 2001; 27(50): 429-433.

[16] Swanbeck G. A new treatment of Ichthyosis and other hyperkeratotic conditions.ActaDerm.Venereol (Stockh). 1968; 48:123-127.

[17] Schupack JL, Haber RS, Stiller MY. The future of topical therapy for cutenous aging. J. Dermatol. Surg. &Oncol.1990; 16:941-944.

[18] Brannon H. Treating wrinkles with alpha hydroxy acid. Dermatol, Surg. 2002; 33:120-142.

[19] Kordel L. Lelordkordel's natural folk remedies. London: Manton Press Ltd; 1976. 177-84p.

[20] Katzung BG. Basic and Clinical Pharmacology. 8th ed. New York: McGrawHill Co; 2001. 1058p.

[21] Quilliot D, Bomam F, Creton C, Pelletier X, Floquet J, Debry G. Phytosterols have an unfavorable effect on bacterial activity and no evident protective effect on colon carcinogenesis. Euro. J. Can. Preven. 2001; 10(3):237-243.

[22] Reynold JEF, editor. Salicylic acid in: Martindale The extra pharmacopoeia. 31st ed. London: The Royal Pharmaceutical Society; 1996. 1093p.

[23] Rios JL, Recio MC, Escandell JM, Anduja I. Inhibition of transcription factors by plant-derived compounds and their implications in inflammation and cancer. Current Pharmaceutical Design. 2009; 15(11):1212-1237.

[24] Thoennissen NH, Iwaski GB, Doan NB. Cucurbitacin b induces apoptosis by inhibition of the JAK/STAT pathway and potentiates antiproliferative effects of gemcitabine on pancreatic cancer cells. Cancer Research. 2009; 69(14):5876-5884.

[25] Lee DH, Iwanski GB, Thoennissen NH. Cucurbitacin: ancient compound shedding new light on cancer treatment. Scientific World Journal.2010; 10:413-418.

[26] Kumar D, Kumar D, Kumar S, Singh J, Rashmi N, Vashistha BD, Singh N. Free radical scavenging and analgesic activities of *Cucumissativus L.* fruit extract. Journal of Young Pharmacist. 2010; 2(4):365-368.

[27] Abu-Reidah I, Arraez-Roman D, Quirantes-Pine R, Fernandez-Arroyo S, Segura-Carretero A, Fernandez-Gutierrez A. HPLC-ESI-Q-TOF-MS for a comprehensive characterization of bioactive phenolic compounds in cucumber whole fruit extract. Food Research International.2012; 46:108-117.

[28] Murad H, Lange D. The Murad Method. New York. Macmillan. 2003.

[29] Murad H, Nyo MA. Evaluating the potential benefits of cucumbers for improved health and skin care.Journal of Aging Research and Clinical Practice. 2016; 5(3):139-141.

[30] Roman-Ramos R, Flores-Saenz JL, Alarcon-Aguilar FJ. Anti-hyperglycemic effect of some edible plants.J Ethnopharmacol. 1995; 48(1):25-32. doi: 10.1016/0378-8741(95)01279-m.

[31] Kaushik U, Aeri V, Mir SR. Cucurbitacins – An insight into medicinal leads from nature Pharmacogn Rev. 2015; 9(17): 12-18. doi: 10.4103/0973-7847.156314.

[32] Alghasham AA. Cucurbitacins - a promising target for cancer therapy. Int J Health Sci (Qassim). 2013;7(1):77-89. doi:10.12816/0006025.

[33] Milder IEJ, Arts ICW, van de Putte B, Venema DP, Hollman PCH. Lignan contents of Dutch plant foods: a database including lariciresinol, pinoresinol, secoisolariciresinol and matairesinol. British Journal of Nutrition.2005; 93:393-402.

[34] Nema NK, Maity N, Sarkar B, Mukherjee PK. *Cucumissativus* fruit-potential antioxidant, anti-hyaluronidase, and anti-elastaseagent. Archives of Dermatological Research. 2011; 303(4):247-252.

[35] Yang Q, Liu T, Kuklina EV, Flanders WD, Hong Y, Gillespie C, Chang MH, Gwinn M, Dowling N, Khoury MJ, Hu FB. Sodium and potassium intake and mortality among US adults: prospective data from the Third National Health and Nutrition Examination Survey. Arch Intern Med. 2011;171(13):1183-1191. doi: 10.1001/archinternmed.2011.257.

[36] Zhao B, Hu L, Dong Y, Xu J, Wei Y, Yu D, Xu J, Zhang W. The Effect of

Magnesium Intake on Stroke Incidence: A Systematic Review and Meta-Analysis With Trial Sequential Analysis. Front Neurol. 2019;10:852. doi: 10.3389/fneur.2019.00852.

[37] Esterbauer H. Cytotoxicity and genotoxicity of lipid-oxidation products. Am J ClinNutr. 1993;57(5 Suppl):779S–785S; discussion 785S–786S. doi: 10.1093/ajcn/57.5.779S.

[38] Tannin-Spitz T, Bergman M, Grossman S. Cucurbitacinglucosides: antioxidant and free-radical scavenging activities. BiochemBiophys Res Commun. 2007; 364(1):181-186. doi:10.1016/j.bbrc.2007.09.075.

[39] Saba AB, Oridupa AO. Search for a novel antioxidant, anti-inflammatory/analgesic or anti-proliferative drug: Cucurbitacins hold the ace. J Med Plants Res. 2010;4:2821-2826.

[40] Nagy IZ. The membrane hypothesis of aging.Boca Raton, FL, CRC Press. 1994. 3-10p.

[41] Onimisi AO, Ovansa JU. Comparative studies on nutritional values of four varieties of cucumber. In: International Conference on Latest Trends in Food, Biological & Ecological Sciences (ICLTFBE'15) October 2015; Dubai (UAE); 2015. p. 38-39. DOI: http://dx.doi.org/10.17758/IAAST.A1015056.

[42] Rios JL. Effects of triterpenes on the immune system.Journal of Ethnophamacology. 2010; 128(1):1-14.

[43] Schwalfenberg GK. The alkaline diet: is there evidence that an alkaline pH diet benefits health? Journal of Environment and Public Health.20122012:727630.

Section 3

Production and Market

Chapter 5

Market Performance and Structure of Cucumber in Ibadan, Oyo State

Iyabo Bosede Adeoye

Abstract

Cucumber is an important vegetable due to its numerous health benefits. There are a number of empirical studies on the economics of production of the commodity, but there is insufficient information on marketing of the commodity. The study was therefore carried out to examine market performance, structure and constraints in cucumber marketing. Primary data was collected from 70 randomly selected actors in the supply chain (54 Retailers and 16 wholesalers). Data collected was analyzed using descriptive statistics, net marketing margin, efficiency and Gini coefficient. Findings revealed that most of the retailers (70.4%) and wholesalers (81.3%) were male. Most of the retailers were within 31–40 years whereas majority of the wholesalers were within 41–50 years and all the marketers had secondary level of education (47.2%). Cucumber marketing was profitable at the wholesale and retail level both at the peak and lean season of cucumber production. Retailers sell an average of 159.8 kg and 83.8 kg weekly in the peak and lean season. Whereas wholesalers sell an average of 1,000 kg and 870 kg weekly in the peak and lean season respectively. Net margin at retail level was higher in the lean season (68.8/kg) compared to the peak season (46.6/kg). Similarly, at wholesale level, net marketing margin at the lean season was 17.5/kg and was higher than 6.3/kg obtained during the peak season. Marketing efficiency was greater than one for the wholesalers and retailers in both seasons. Gini coefficient of retailer was 0.32 and 0.36 for the peak and lean season indicating that the market was competitive. There was inequality in the wholesale market as indicated by the gini coefficient result. The most important constraint to cucumber marketing was perishability of the produce and price fluctuation. The study recommends improved sensitization on adequate post-harvest handling practices and storage to reduce the levels and consequences of the perishable nature of the commodity.

Keywords: Cucumber, market performance, market structure, constraints

1. Introduction

Cucumber is one of the most important exotic vegetables in Nigeria [1] and are valuable source of antioxidant nutrients including vitamin C, beta carotene and Manganese with about 95% water [2]. The crop is important in prevention of human constipation and improvement in digestion [3]. Aside the nutritional benefit of cucumber, production of the commodity is also profitable [2]. Marketing is crucial to link production and consumption and facilitate increased agricultural productivity and employment [4]. Thus, efficient marketing system is germane to improve market organization in order to satisfy consumers' needs and wants [4]. Market performance is the ultimate result derived from the market which include outcome from various market activities [5]. Furthermore, profitability is a key elements of financial performance [6]. Market structure is the characteristics of the organization of the market that influence the nature of competition and pricing within the market [7]. Market structure for agricultural products in Nigeria is not perfectly competitive due to collusive tendencies of sellers [8]. It is worthy of note that analysis of market structure, efficiency of an agricultural product determines whether the market is a perfectly competitive, oligopolistic or monopolistic market [9].

Most socioeconomic research on cucumber in Nigeria have focused on economics of production [1, 2]. This includes analysis of profitability and major constraints of cucumber production in two local government areas of Rivers state [2]. They found that cucumber production was profitable and the major constraints in cucumber production in the study area were pest and disease attacks, poor funding and unfavorable climatic conditions. Profitability and efficiency of cucumber production in Iseyin Local Government area of Oyo State was also evaluated [1]. They found that cucumber production was profitable in the study area and they found that Age, Education status of farmers and access to credit were the factors affecting technical efficiency of the farmers in cucumber production in the study area. Factors affecting cucumber farmers' market participation were examined in Odukpani Local Government Area of Cross River State, Nigeria [10]. They found that distance to the market, market information and quantity of cucumber harvested were significant and important factors affecting the ability of the smallholder farmers to participate in the output market.

Past empirical studies on Market structure and performance of vegetables were on watermelon. Watermelon marketing was analyzed in Oyo State, Nigeria and the study revealed that watermelon marketing is profitable and efficient [11]. They also found that there was price discrimination and product differentiation in the market based on size and quality of the product. Structure and performance of pumpkin marketing was analyzed in Nassarawa State [4] and the study revealed that pumpkin marketing was profitable and the market was noncompetitive. Thus, there is dearth of information on market structure and performance of cucumber in Oyo state, Nigeria. The study intends to fill the above information gap.

2. Methodology

The study area was Ibadan ($7^{o}23^{1}47^{1}$N $3^{o}55^{1}0^{1}$E), capital of Oyo state. The city has a tropical wet and dry climate. There are eleven local government areas in Ibadan. The city is a major center for trade in horticultural commodities such as plantain, banana, cucumber, watermelon and arable crops [12].

Multistage sampling technique was employed in the selection of respondents. The first stage involved purposive selection of two local government areas (Ibadan North and Ibadan North West) with effective cucumber marketing while stage two involve selection of major markets noted for cucumber marketing in the local government while the third stage involve random selection of 54 retailers and 16 wholesalers in the selected markets.

Primary data was collected with the aid of well-structured questionnaire. Socio economics characteristics of marketers was analyzed using descriptive statistics such as frequency and percentages. Market performance was analyzed using the net marketing margin and efficiency while the market structure was estimated using Gini coefficient. Gini coefficient is used to measure statistical dispersion and is a concentration indices and it utilizes market shares to determine the extent of market concentration [13]. The close the value of Gini coefficient is to unity, the greater the degree of inequality [14]. The net marketing margin is the difference between gross return and total cost involved in marketing while the marketing efficiency is used to compare return per unit cost.

3. Results and discussion

3.1 Socioeconomic characteristics of marketers

Findings revealed that most of the retailers (70.37%) and wholesalers (81.25%) were male indicating dominance of male folks in cucumber marketing in the study area. Most of the retailers (42.59%) were within 31–40 years age group whereas majority of the wholesalers (43.75%) were within the age group of 41–50 years indicating that the wholesalers were older than the retailers in the study area. Most of watermelon marketers which is also a curcubit like cucumber were also within the age group of 31–40 years indicating that the marketers were young and agile [11]. All the wholesalers were married while 66.67% of the retailers were married. Most of the retailers (74.0%) and wholesalers (68.8%) were educated at the primary and secondary level. Whereas most of the watermelon marketers in Oyo state (54.4%) had no formal education [11]. Most of the retailers had up to 6–10 years' (44.44%) experience in cucumber marketing whereas the wholesalers had more years of experience in the marketing 11–15 years (31.25%). Most of the retailers (87%) and wholesalers (75%) had 1–5 members in their respective household. All the wholesalers had marketing of vegetable as their main occupation while 98.1% of the retailers had marketing as their main occupation and the major source of capital was from personal savings. Only 22% of the retailers were members of association while majority of the wholesalers (75%) were members of association such as cooperative society. Majority of the retailers (88.9%) and Wholesalers (56.3%) have not received training on vegetable marketing. The marketers had poor extension contact in the study area. Majority of the retailers (98.1%) and wholesalers (93.8%) had no contact with extension agents in their marketing activities (**Table 1**).

Characteristics		Retailers	Wholesalers	Average %
Sex	Male	70.37	81.25	75.81
	Female	29.63	18.75	24.19
Age	Below 20			
	21–30	25.93		12.97
	31–40	42.59	18.75	30.67
	41–50	20.37	43.75	32.06
	51–60	3.70	31.25	17.48
	61–70	3.70	6.25	4.98
	Choose not to say	3.70		1.85
Marital Status	Single	31.48		15.74
	Married	66.67	100	83.34
	Divorced	1.85		0.93
Educational level	No formal	25.9	31.3	28.6
	Primary	29.6	18.8	24.2

Characteristics		Retailers	Wholesalers	Average %
	Secondary	44.4	50	47.2
	Tertiary			
Years of Experience	1–5	12.96	18.75	15.86
	6–10	44.44	6.25	25.35
	11–15	16.67	31.25	23.96
	16–20	11.11	25.0	18.06
	21–25	11.11	12.50	11.81
	Above 26	3.7	6.25	4.98
Household Size	1–5	87.0	75	81.0
	6–10	13.0	25	19.0
	11–15			
	16–20			
Main occupation	Marketing of vegetables	98.1	100	99.1
	Farming	1.9		1.9
	Civil Servants			
Major Source of capital				
	Personal savings	94.4	100	97.2
	Friends and relatives	5.6		2.8
	Cooperatives			
	Bank loans			
Members of Association				
	Yes	22.2	75	48.6
	No	77.8	25	51.4
Vegetable training	Yes	11.11	43.75	27.43
	No	88.89	56.25	72.57
Receive Agricultural Extension visit				
	Yes	1.9	6.3	4.1
	No	98.1	93.8	95.9
Source: Field Survey, 2020.				

Table 1.
Socioeconomic characteristics of the marketers.

4. Market performance of cucumber

Although there are peak and lean season of marketing the commodity. Cucumber marketing is throughout the year. Market performance of cucumber was evaluated using net marketing margin and efficiency. Average quantity sold by retailers weekly in the peak and lean season were 159.8 kg and 83.8 kg respectively. This indicates that more quantity were sold in the peak season compared to the lean

Variable	Retail level	
	Peak period Cost/kg	Lean period Cost/Kg
Average quantity sold (kg) weekly	**159.8**	**83.8**
Average cost price/kg	N64.3	162.5
Average marketing cost/kg	18.68	18.68
Total cost	82.98	181.18
Average selling price/kg	129.6	250
Net margin/Kg	46.6	68.82
Marketing Efficiency	1.6	1.4
	Wholesale level	
Average quantity sold (kg) weekly	**1,000**	**870**
Average cost price N/kg	43	130
Average marketing cost/kg	15	15
Total cost	58	145
Average selling price/kg	64.3	162.5
Net margin/Kg	6.3	17.5
Marketing Efficiency	1.1	1.1

1 $ = 395.22 Nigerian naira.
Source: Field Survey, 2020.

Table 2.
Marketing margin and efficiency of cucumber.

season of production of the commodity. Average cost price per kg of cucumber in the peak season was N64.3 against N150/kg that was obtained in the lean season. The total cost incurred per kilogram was N82.98 while the total revenue was N129.6 in the peak season. The marketing margin and efficiency obtained during the peak season were N46.6/kg with the marketing efficiency of 1.6. This indicated that cucumber marketing is profitable (**Table 2**).

Similarly in the lean season of cucumber production, average cost incurred in the retail marketing was N181.18/kg with the total revenue of N250/kg. The marketing margin obtained in the lean season was N68.82/kg with the marketing efficiency of 1.4 indicating profitability of marketing cucumber in the study area. Although marketing efficiency was higher in the peak season compared to the lean, may be attributable to the lower marketing cost in the peak season (**Table 2**).

For the wholesalers, average quantity of cucumber sold during the peak season of cucumber production was 1,000 kg while average of 870 kg was sold during the lean season. Average cost incurred during marketing of cucumber per kilogram was N58 in peak season against N148 that was obtained in the lean season. Marketing margin and efficiency of cucumber in the peak season were N6.3k per kg and 1.1 indicating that cucumber marketing at wholesale was profitable. Marketing margin and efficiency during lean period was N17.5/kg while the marketing efficiency was 1.1 indicating efficient marketing system (**Table 2**).

4.1 Market structure of cucumber (retail)

Cucumber market structure was estimated using Gini coefficient. Results revealed that marketers within the sale ranges of N5001- N25, 000 constituted

Sales N			**Peak season**				
	Frequency	**% of retailer ∑X**	**Cumulative % of retailer**	**Total value of sales**	**% of total sales**	**Cumulative Y**	**∑XY**
<5,000	0						
5,001 – 25,000	34	63	63	533,952	49.5	49.5	0.31
25,001 – 45,000	20	37	100	544,320	50.5	100	0.37
Total							**0.68**
Gini Coefficient	**0.32**						
			Lean season				
<5,000	2	3.7	3.7	9,100	0.91	0.91	0.0
5,001 – 25,000	35	64.8	68.5	527,800	53.2	54.11	0.35
25,001 – 45,000	15	27.8	96.3	364,000	36.7	90.81	0.25
45,001 – 65,000	2	3.7	100	91,000	9.2	100	0.037
Total	54	100		991,900			**0.64**
Gini Coefficient	**0.36**						

Source: Field Survey, 2020.

Table 3.
Gini coefficient of retail cucumber marketing.

63% of retailers and this accounted for 49.5% of total sales of cucumber during the peak season (**Table 3**). Similar trend was obtained in the lean season, marketers within the sale range of 5,001–25,000 also constituted 64.8% of retailers and this accounted for 53.2% of cucumber total sales in the lean season. Gini coefficients for the peak and lean season were 0.32 and 0.36 respectively. This indicated that cucumber marketing at retail level in the peak and lean season was competitive. This indicated that there are many retailers in the market and they will not be able to influence price by increasing or decreasing quantity supplied to the market. This also indicate adequate equality in the market.

4.2 Market structure of cucumber (wholesale)

Findings revealed that wholesale within the sales range of 25,001–45,000 and 45,001–65,000 constituted the greatest percentage of the wholesaler and this accounted for 13.8 and 25% of total sales. Gini coefficient for the wholesaler during the peak and lean season were 0.5 and 0.57 respectively. This implied inequality in quantity of cucumber sold among the wholesaler and sales is concentrated in the hand of few marketers and cucumber marketing at wholesale level was noncompetitive in the study area. Similarly, high inefficiency was also observed in the market structure of watermelon at Akure [14], where it was found that the Gini coefficient of watermelon was 0.7318 indicating inequality in the market (**Table 4**).

4.3 Constraints in cucumber marketing

The most important constraint to cucumber marketing was perishability of the commodity. Vegetables are highly perishable and sensitive to harvest and post-harvest handling systems. It was observed during the survey that most of the marketers use bags in packaging of the crop with attendant physical damages and

Sales N	Peak season						
	Frequency	**% of wholesaler ∑X**	**Cumulative % of wholesaler**	**Total value of sales**	**% of total sales**	**Cumulative Y**	**∑XY**
25,001 – 45,000	5	31.3	31.3	141,460	13.8	13.8	0.043
45,001 – 65,000	5	31.3	62.6	257,200	25.0	38.8	0.12
65,001 – 85,000	1	6.3	68.9	77,160	7.5	46.3	0.03
85,001 – 105,000	3	18.8	87.7	308,640	30.0	76.3	0.14
105,001-125,000	1	6.3	94.0	115,740	11.3	87.6	0.06
125,001-145,000	1	6.3	100	128,600	12.5	100	0.1
Total				1,028,800			0.5
Gini Coefficient	0.5						
			LEAN SEASON				
25,001 – 45,000	2	12.5	12.5	71,500	3.2	3.2	0.004
45,001 – 65,000	4	25	37.5	240,500	10.6	13.8	0.0345
65,001 – 85,000	0	0	37.5	0	0	13.8	0
85,001 – 105,000	2	12.5	50	195,000	8.6	22.4	0.028
105,001-125,000	0	0	50	0	0	22.4	0
125,001-145,000	0	0	50	0	0	22.4	0
145,001-165,000	0	0	50	0	0	22.4	0
165,001-185,000	0	0	50	0	0	22.4	0
185,001-205,000	4	25	75	780,000	34.5	56.9	0.14225
205,001-225,000	0	0	75	0	0	56.9	0
225,001-245,000	2	12.5	87.5	455,000	20.1	77.0	0.096
245,001-265,000	2	12.5	100	520,000	23.0	100	0.125
Total				2,262,000			0.43
265,001-285,000	GINNI COEFFICIENT	**0.57**					

Table 4.
Gini coefficient of wholesale cucumber marketing.

	Yes	**No**
Perishability	69(98.6)	1(1.4)
High cost of transportation	64(91.4)	6(8.6)
Price fluctuation	67(95.7)	3(4.3)
Inadequate capital	67(95.7)	3(4.3)
Storage Problem	67(95.7)	3(4.3)
Poor marketing information	49(70.0)	21(30.0)

Source: Field survey, 2020.

Table 5.
Constraints in cucumber marketing.

losses. Similarly, high perishability is one of the constraints in cucumber production in Southeast Nigeria [15]. Price fluctuation is rampant and is always due to fluctuation in the supply of the commodity. Other constraints were inadequate capital and storage problem, high cost of transportation and poor marketing information (**Table 5**).

5. Conclusion

Cucumber marketing was dominated by the male folks at both retail and wholesale level. Most of the retailers were within the age group of 31–40 years and were younger than the wholesaler. Cucumber marketing at both retail and wholesale was profitable in both the peak and lean season. The analysis of the market structure at the retail level showed adequate equality while market structure at the wholesale level reflect inequality among the actors. The major constraints in cucumber marketing were perishable nature of the produce, price fluctuation and storage problem. The study advocates for the use of improved packaging in the marketing of the commodity and enhanced sensitization on the importance of adequate post-harvest handling and proper storage to reduce the problem of perishability encountered in the commodity supply chain.

Acknowledgements

I wish to acknowledge Mr. Habib Ganiyu for assisting in data collection for the study.

Questionnaire

Market Performance and Structure of Cucumber in Ibadan, Oyo State

Questionnaire code ____________ Date of interview:____________
Phone no ....

a. State..

b. Local Government Area...

c. Name of Market ....

Section A: Socioeconomic characteristics of the Marketers

1. Sex 1. Male 2. Female

2. Age (Years): ....

3. Marital status 1. Single 2. Married 3. Divorced 4. Others please specify

4. Educational level 1. No formal 2. Primary 3. Secondary 4. Tertiary 5. Others please specify

5. Years of experience in marketing ....

6. Type of marketer a. Wholesaler Yes...... No..............

   b. Retailer Yes...... No............

7. Household size ....

8. What is your main occupation 1. Trading of cucumber 2. Farming 3. Civil servants 4. Others please specify.

9. Major source of capital i. Personal savings ii. Friends and relatives iii. Cooperative loansiv. Bank loans

10. Are you a member of any association 1 Yes... 2. No

11. Have you ever attended any training on vegetable marketing 1. Yes 2. no

12. Do you receive agricultural extension services? 1. Yes 2. No

Section B: Market performance and structure of Cucumber

13. How many bags (40 kg weight) do you buy in a week: i. peak season .... ii. Lean season ....

14. Cost price/bag: i. Peak season .... ii. Lean season ....

15. Loading cost per bag in i. Peak season .... ii. Lean season ....

16. Transport cost/bag i. Peak season .... ii. Lean season ....

17. Average cost of stall per month ....

18. Selling price per bag i. Peak season .... ii. Lean season ....

Section C: Constraints in marketing cucumber

S/N	Constraints	Yes	No
1	Perishability		
2	High cost of transportation		
3	Price fluctuation (seasonality)		
4	Inadequate capital		
5	Storage problem		
6	Poor marketing information		

Any comment ....

Author details

Iyabo Bosede Adeoye
National Horticultural Research Institute, Ibadan, Nigeria

*Address all correspondence to: bosedeadeoye2005@gmail.com

References

[1] Adeoye IB and Balogun OL. Profitability and Efficiency of Cucumber Production among Smallholder Farmers in Oyo State, Nigeria. Journal of Agricultural Sciences, 2016; 61(4): 387-398.

[2] Elum ZA, Etowa, EB and Ogonda AU. Economics of Cucumber Production in Rivers State, Nigeria. Agro-Science Journal of Tropical Agriculture, Food, Environment and Extension, 2016; 15(2): 48-53.

[3] Lutfa A, Happy FA and Yeasmin F. Production process and marketing system of Cucumber: A Socioeconomic Study In Mymensingh District Of Bangladesh, SAARC J. Agric., *2019; 17(1): 135-147*

[4] Girei AA, Kasali R, Ogezi E and Nnodu ON. Analysis of Structure and Performance of Pumpkin Marketing in Nasarawa State, Nigeria. Direct Research Journal of Agriculture and Food Science, 2018; 6(7): 166-172.

[5] Onyango CO. Analysis of Structure, Conduct and Performance of small ruminant stock market participants of Isiolo Nairobi trading market, Kenya M.Sc Project. Agricultural and Applied Econoics of Egerton University, 2013; xiii+93pp.

[6] Azim MD, Helaluddin A, and Shibbir Khan ATM. Operational Performance and Profitability: an empirical study on the Bangladeshi Ceramic Companies. International Journal of entrepreneurship and Development Studies, 2015; 3(1), 63-73.

[7] Olukosi JO, Isitor SU and Ode MO. Introduction to Agricultural marketing and Prices: Principle and Application, Living books Series, G.U. Publications, Abuja, Nigeria, 2005.

[8] Imoudu PB and Afolabi JA. An Assessment of the performance of plantain marketing in Ondo State, Nigeria. Journal of Applied Science, 2002; 5:2690-2697

[9] Adeoye IB and Ibe RB. Market Structure and Performance of Fresh Tomatoes in Ibadan Metropolis, Oyo State, 2013, Proc. 2nd All Africa Horticulture Congress Eds.: K. Hannweg and M. Penter, Acta Hort. 1007.

[10] Ohen SB, Umeze GE, Cobham ME. Determinants of Market Participation by Cucumber Farmers in Odukpani Local Government Area, Cross River State, Nigeria, Journal of Economics and Sustainable Development, 2014; 5(2): 188-196.

[11] Kasali R, Aremu FJ and Shittu BA. An Economic Analysis of Watermelon Marketing in Oyo State, Nigeria. Production Agriculture and Technology, 2015; 11(2):43-52.

[12] Wikipedia . Ibadan. https://en.wikipedia.org/wiki/Ibadan, 2021.

[13] Garba ID, Sanni SA and Adebayo CO. Analyzing the structure and performance of Shea butter market in Bosso and Borgu Local Government Areas of Niger State, Nigeria, International Journal of Service and Technology, 2015; 8(2):321-336.

[14] Oseni JO. An appraisal of the structure and conduct of watermelon marketing in Akure Metropolis, Ondo State, Nigeria. Sky Journal of Agricultural Research, 2015; 4(4):080-089.

[15] Umeh OA, and Ojiako FO Limitations of Cucumber (Cucumis Sativus L) Production For Nutrition Security In Southeast Nigeria, INT'L JOURNAL OF AGRIC. AND RURAL DEV, 2018; 21(1): 3437-3443,

Section 4

Cultivation and Management

Chapter 6

Growth Promoting and Stabilizing of Cucumber Plants Cultivated in Soilless Cultivation Systems Using Biostimulators

Michael Henry Böhme

Abstract

Cultivation of cucumber in greenhouses is predominantly carried out in soilless cultivation systems mainly in substrate culture. The use of organic and completely compostable substrates is of rising interest in such systems, unclean sheep wool was formed as substrate slabs, peat slabs and coconut fiber slabs were compared with mineral substrates rockwool and perlite. In general can be stated, intensively in greenhouses cultivated crops such as cucumbers, suffer often from inadequate abiotic or biotic growth conditions in particular in the rhizosphere. Many studies were done to find growth promoters or biostimulators to stabilize the growing condition in the rhizosphere, in case of stress situation as fluctuating salt concentration EC and pH value, but also in case of temperature stress. K-Humate, Lactate and *Bacillus subtilis* were investigated as biostimulators in such situations. Different concentration and combination of these biostimulators were investigated but also the methods of application on leaves and roots respectively. Very successful was used for the stabilization of the EC value for cucumber plants growing in substrate the application of K-Humate and *B. subtilis* (FZB24®) as single component and combined. Following the results, it can be assumed, the application of the combined biostimulators with all substances if applied over the roots was a stimulating effect visible, whereas if applied over the leaves an inhibiting effect for cucumber plant growth. Application of the biostimulators solution with all three components BS-FZB24® (0.2%), K-Humate (0.01%), and LACTOFOL "O" (0.1%) were tested regarding their effects in case of strong but short time pH and temperature stress the growth of cucumber plants. There are a strong correlation between green biomass of treated cucumber plants and their root mass. It can be assumed that one of the effects of stress prevention through the biostimulators is primarily based on increasing root growth.

Keywords: soilless cultivation, organic and mineral substrates, humates, lactates, *Bacillus subtilis*, abiotic stress influence

1. Introduction

Cucumbers (*Cucumis sativus* L.) is a vegetable originated from the rain forest of northeastern India. Therefore, they are growing in particular in temperate areas in glass-greenhouses, in order to secure the appropriate temperature and humidity whereas in sub-tropical regions mostly in plastic greenhouses to secure the suitable humidity or to safe the cucumbers from heavy rainfalls. In all cases very important are the growing conditions in the rhizosphere. In the rainforest, the cucumbers are growing in a soil with a high amount of organic matter. Therefore, substrates suitable for cucumber cultivation in greenhouse should have specific physical properties, as pore volume, air and water capacity and density of the substrates [1]. In general, substrates from three groups can be used, mineral or inert, artificial and organic (**Table 1**).

Which of the substrates is suitable, depends also from availability in the region of the basic material, the economically preparation and environmental tasks as possibilities of reuse or environmentally friendly waste disposal. All substrates have some limitations in comparison to the natural growing conditions of cucumbers:

- Strong reduction of the volume for root growth
- Different physical, chemical and biological conditions in the substrates
- Missing of a buffer for stress situation as fluctuation of the pH and salt concentration (EC) or limitations of macro- or micro nutrients
- Under natural soil or organic growing media conditions, humates are available as a buffer for fluctuation of salt concentration
- Microorganisms in the rhizosphere can have different functions as mineralization of organic material, stress reduction and as growth regulator.

A proper regulation of the growing condition in the rhizosphere of cucumbers in greenhouses, is also from high importance because the wide relation of shoot and leave to the roots - 100:1, that means a comparable little root system have to secure water and nutrients for a big biomass.

Nowadays, mainly mineral or Rockwool is used as substrate in greenhouses for fruity vegetable production as tomatoes and cucumbers. Therefore, most of the studies with different substrates for cucumber cultivations are including as 'control' the Rockwool as slabs or as granules in containers.

Mineral - inert	Organic	Artificial
Expanded Clay	Coconut fiber	Aggrofoam*
Gravel	Wood fiber	PU-foam
Perlite	Peat	Polystyren-foam
Pumice	Mix substrate	Polyphenol-foam
Rockwool	Compost	UMF-foam
Sand	Moss peat	
Zeolite	Straw	
	Sheepwool	
	Pine bark	

PU = Polyurethan; UMF = Urea-Methanal foam.
**PU foam mixed with recycled PU-granules.*

Table 1.
Classification of substrates for hydroponically cultivation of vegetables as cucumbers and others.

In general can be stated, intensively in greenhouses cultivated crops such as cucumbers, suffer often from inadequate abiotic or biotic growth conditions in particular in the rhizosphere. Many studies were done to find growth promoters or biostimulators to stabilize the production process or to enhance plant growth of cucumbers under these conditions. Following such reports and own research the wide range of effects produced by Humates based on humic acids, have beneficial effects on the growing conditions in the rhizosphere. Many experiments were showing positive effects of different Humates as 'Bioregulator' in substrate culture of cucumbers. Other organic substances with a similar effect were tested as Lactates (salts of lactic acid) because these substances have proven biostimulatory effects and as an approach to improve the nutrient balance and plant vitality. Investigations have shown that Lactates have more stable bonds with several metal ions than other chelates. This effect can be very important to improve the nutrient supply of the cucumbers in soilless cultivation systems.

Furthermore, as mentioned the biological conditions in these intensive cultivation systems are completely changed, in particular in the inert and artificial substrates. Many microorganisms from the rhizosphere can influence plant growth and plant health positively, and are therefore often referred to as "plant growth promoting rhizobacteria" [2]. Different microorganisms are tested regarding these effects best result could be induced by *Bacillus subtilis* (syn. *B. amyloliquefaciens ssp. plantarum*) regarding these growth and plant health promoting mechanisms as well as the interactions between them.

In several studies, cucumber growth in different substrates was investigated using mineral nutrient solution and with organic nutrients. Furthermore, plant growth-promoting agencies as the mentioned Humates, Lactates and Bacillus subtilis were investigated single or in combination regarding their effect as Bioregulator in the rhizosphere as well on the epigean part of the plants.

Following these investigations, different effects in relation to plant growth factors or conditions could find out in particular under stress conditions for plant growth (**Figure 1**):

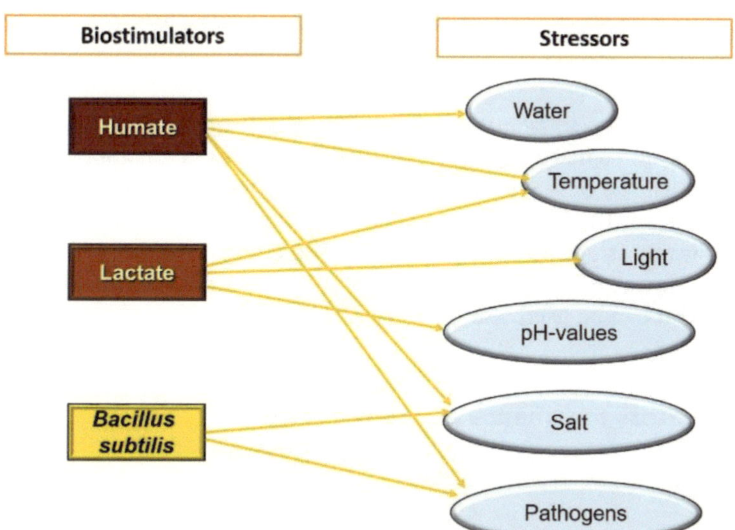

Figure 1.
Effect of biostimulators humate, lactate and Bacillus subtilis *on stressors during growth of cucumbers under protected cultivation.*

2. Cultivation systems for cucumber growing in greenhouses

In the last 45 years different hydroponic systems are developed in the world, the different systems can be subdivide in four groups [3]. Whereas, cucumbers are cultivated mostly in the first and second system.

2.1 Substrate culture

Hydroponic systems in a small amount of substrate (3 to 15 l per plant) filled in containers, bags, slabs or channels. Slabs of rockwool, peat or coconut fiber are used. This system is combined with different types of trickle irrigation. There are 'open systems' in which the drainwater flows out of the substrate slabs or cultivation container in the ground. Preferable used are the 'closed systems' in which the growing medium is lying in channels and the nutrient solution is recirculating.

2.2 Water culture

The plants are cultivated without any substrate, except the substrate for propagation of the young plants. The plants are staying or hanging also in channels in recirculating nutrient solution. Quite known systems are the 'Nutrient-Film-Techniques' (NFT) and the 'Deep-Flow-Techniques' (DFT). Also gravel culture can be grouped belong to the water culture, because the gravel has only the function to stabilize the root system.

2.3 Aeroponics

The plants are cultivated without any substrate and staying or hanging on plastic-foam-sheets and the root system is hanging in a closed space. The nutrient solution is given by very fine nozzles as a kind of fog-system. Important is a high frequency of spraying with nutrient solution, advisably to spray every 2 to 4 minutes for 20 to 30 seconds according to the developmental stage of plants. This system is rarely used for cucumber cultivation.

2.4 Aquaponics

Is a combination of aquaculture with typical fish keeping in ponds and the water culture system 'nutrient-film-technique' (NFT) as a closed aquaponics, where the plants were fertilized only with the fish water. This technique becomes more and more important for the food production and has an increasing popularity, even though the system is not methodologically sound and completely scientifically investigated. Nowadays, this system is mainly used to cultivate tomatoes and leafy vegetables [4] it can be used also for cucumber cultivation.

3. Biostimulators to enhance cucumber growth

3.1 Lactate

According to previous investigations, lactates (salts of lactic acid) seem to produce bioregulatory effects. The application of lactates was tested as an approach to improve the nutrient balance and plant vitality. Investigations have shown that lactates have more stable bonds with several metal ions than other chelates.

Therefore, substances have been used as fertilizers and as bioregulators. Lactates are available under the brand name LACTOFOL® [5, 6]. That suspension fertilizer was designed mainly for foliar application and patented as a plant growth and development regulator. The composition of LACTOFOL O® is shown in **Table 2**.

3.2 Humates

Humates are the salts of Humic acids which are complex polymers including amino acids, amino sugars, peptides, aliphatic compounds involved in linkages between the aromatic groups [7]. There are different types of humates related to the organic material used and the method of extraction of the Humic Acid; in the experiments were used different types:

- Peat used by Merck
    - Na Humate
- Brown coal
    - K Humate
    - NH_4 Humate
- Leonardite G (Germany)
    - Fulvic Acid
    - K Humate
- Leonardite R (Russia)
    - K Humate
- Coconut fiber (Mexico)
    - K Humate

Components	Unit	LACTOFOL O®	Components	Unit	LACTOFOL O®
Lactic acid	%	10	Magnesium	%	0.1
Riboflavin	mg/l	0.5	Iron	%	0.4
Ascorbic acid	mg/l	3	Boron	mg/l	300
Thiamine	mg/l	0.1	Copper	mg/l	200
Nitrogen	%	30	Manganese	mg/l	250
Phosphorus	%	7.5	Zinc	mg/l	125
Potassium	%	15	Molybdenum	mg/l	18
Calcium	%	0.5	Cobalt	mg/l	6

Table 2.
Composition of LACTOFOL O®.

Na Humate is not very useful as Bioregulator, because the negative effects of high Na in the nutrient solution or substrates, therefore K Humates or NH_4 were used in the experiments.

In relation to crop growth or soil condition can be divided in Direct and Indirect effect of humates:

- Direct effects of humates
  - to increase the vitality and stimulating plant growth,
  - to increase the germination activity and accelerate the germination of seeds.
  - to improves the uptake of some nutrients and enhances the transport and availability of micronutrients in the plant
- Indirect effects of humates
  - Regulation of soil properties,
  - to improve thermal conditions and soil or substrate structure,
  - to enhance the ion - exchange capacity,
  - to stimulate development of micro-organisms.

3.3 Microorganisms

Intensively cultivated crops such as cucumbers in soilless cultivation systems are stressed because the growth conditions in the rhizosphere are completely different in contrast to the natural conditions. Therefore, studies were done to improve the growth condition in the rhizosphere, to enhance plant growth and to find growth promoters to stabilize the production process. Investigations were carried out to find effective biostimulators and agencies for plant protection from pest and diseases. Different bacteria species and their strains were investigated and practically used for different plants mainly in protected cultivation [8, 9].

Many microorganisms from the rhizosphere can influence plant growth and plant health positively and in case of bacteria they are often referred as "plant growth-promoting rhizobacteria" However, their effects have to be seen as the complex and as a cumulative result of various interactions between plant, pathogen, antagonists, and environmental factors [2].

Following many investigations in the soil are bacteria present in an average amount of 6 × 108 cells/g of soil, and with a live weight of about 10,000 kg/ha, they are the most common microorganisms in the soil, bacillus species are one of them. The species *Bacillus subtilis* is common in the nature and can be found in every compost pile, but it was important to find active strains promising for developing marketable products. Based on several investigations, some strains of *Bacillus subtilis* are already used in industrially and agricultural fields [10]. An effective strain is in this regard is FZB24® in particular used in agriculture also as growth promoter in the rhizosphere of the cultivated plants [11].

There are various effects induced by *Bacillus subtilis* (syn. *Amyloliquefaciens* ssp. *plantarum*), and different mechanisms of these effects as well as the interactions between them.

Following the investigations in the research and practical experiences in agriculture and horticulture, it was found out that *Bacillus subtilis* could have different effects:

- Growth promoting of plants.
    - Formation of growth hormones and enzymes for nutrient mobilization, based on better this nutrient availability and the nutrient uptake can be enhanced.
    - Improvement of seed and tuber germination.
    - Improvement of root formation of plants.
    - Improvement of plant growth, biomass production and marketable yield.
    - Improving of the earliness of generative development and vegetative growth.
- Promoting of plant health.
    - Improving plant resilience.
    - Reducing the effects of diseases in terms of intensity and frequency.
    - Achieve resistance of the plants in particular against soil born diseases.

4. Mineral, organic and synthetic substrates used in the substrate cultivation system of cucumbers

As mentioned in the introduction different substrates can be used for cucumber cultivation in substrate culture systems. In principle, most substrates are suitable for successful cucumber growing, if adequately supplied with water and plant nutrients.

In one experiment were compared substrates from the three groups (**Table 1**), because the ecological aspects are becoming more and more important when choosing substrates, as well the economic efficiency is important. Substrate were filled in containers or available as mats, between 8 and 9 liters of substrate were available per plant. The aim was to use substrates with different physical and chemical parameters (**Table 3**).

- Rockwool (Grodan) mats as standard substrate,
- Perlite, grain size between 0.6 mm and 1.0 mm diameter.
- Urea-methanal foam,
- Organic substrate, consisted of pine bark (40% v /v), low-bog peat (40% v/v) and manure straw mix (20% v/v).

Comparing the four substrates used, the significantly highest total yields of cucumbers were obtained in the variants 'organic substrate'. The differences in crop

Substrates	Physical characteristics [% v/v]			Chemical characteristics		
	WC	AC	PV	Density [kg/m³]	pH	CEC (mval/1)
Rockwool	54	32	86	80	7.5–8.8	—
Perlite	36	52	88	98	6.5–8.0	5
UM-Foam	45	45	90	25	6.0–6.7	2
Organic Substrate (OM)	50	40	90	350	6.0–6.5	165
Target values *	45–65	20–40	70–95	30–500	6.7–7.0	50

WC = water capacity; AC = air capacity; PV = pore volume; CEC = cation exchange capacity.
**[1].*

Table 3.
Physical and chemical characteristics of substrates used in experiments with cucumber growing in substrate culture.

Substrate	Months of harvest						Average
	February	March	April	May	June	July	
Rockwool	0.40 b	4.95 b	6.90 a	7.78 a	6.80 bc	5.77 b	5.43
Perlite	1.45 a	4.78 b	6.63 ab	6.54 ab	7.10 b	8.48 a	5.83
UM-Foam	1.10 a	4.30 b	6.10 ab	5.65 b	6.60 c	8.05 a	5.30
Organic Subst.	1.37 a	6.24 a	7.36 a	7.14 a	9.25 a	7.27 a	6.44

Different letters indicate significant differences, LSD, $p \leq 0.05$.

Table 4.
Yield of greenhouse cucumber in mineral, artificial and organic substrate (kg/m²).

yield of the investigated substrates in the examined months (**Table 4**) can be due, because different growing conditions, as:

- lower substrate temperatures and lower water holding capacities in urea foam and perlite
- higher temperatures and better sorption capacity in the organic substrate

The development of the leaf area of cucumber plants was examined for the variants with rockwool mats, urea foam and perlite. The following leaf areas were determined as the mean of all measuring dates: in rockwool 8215 cm^2, in urea foam 7889 cm^2, and in perlite 7438 cm^2.

5. Effects of biostimulators on the growth and cucumber plant development

5.1 Mineral and organic substrates used in substrate culture of cucumbers treated with biostimulators

In Europe, about 87 million sheep are produced, in Germany about 2.2 million. Often, there is a lack of capacity for cleaning the sheep wool, so unclean sheep wool is available as waste material.

The objective of this experiment was to investigate the development and yield of cucumber grown on sheep wool slabs in comparison with peat and coconut fiber slabs as well as Rockwool slabs and perlite in containers. Different sheep wool slabs in size and added components were tested, the highest stability was obtained with sheep wool slabs in combination with coconut fibers.

In this experiment following questions were investigated:

- Does sheep wool slabs have appropriate physical parameters for cucumber cultivation and how many months they can be used?
- How do the physical properties change in comparison with other organic and mineral substrates?
- How do biostimulators influence cucumber plant growth in the investigated substrates?

Sheep wool slabs used in this experiment were compared with the following substrates [12]:

- Coconut fiber (width 25 cm, length 100 cm, height after wetting 9 cm) from the company 'Dutch Plantin'.
- Peat slabs 'GroBoard' (width 20 cm, length 100 cm, height after wetting 9 cm) from the KEKKILÄ Oyj company Finland.
- Perlite, average grain size between 0.06 mm and 1.5 mm, properties see **Table 3** was filled in Container with 7 L volume.
- Rockwool slabs (width 20 cm, length 100 cm, height 7.5 cm) from the company 'Pargro' from Finland.

In this experiment fifty percent of the cucumber plants of each substrate variant were treated three times with a biostimulator solution containing 0.08% K-Humate (Fa. Humintech), 0.2% Lactofol (O) (Fa. ECOFOL) and a spore suspension (10^7 cfu ml^{-1}) of *Bacillus subtilis* FZB 24®. To the plants were applied 20 ml of this solution three times in weekly intervals starting with first treatment in 6–7 leaves stage, after transplanting. Plants of *Cucumis sativus* L. 'Indira' were used for the experiment. The experiment was conducted during two cultivation periods, first from November until April and second from June until November in the next year.

5.1.1 Effects of biostimulators growing cucumbers in mineral and organic substrates

For the soilless cultivation of the cucumber plants substrates was selected with different physical properties (**Table 5**). In general substrates for cucumbers should have high air capacity with a range between 20 to 40%, during the cultivation time often the air capacity is decreasing very much and water capacity increasing, therefore a certain stability is necessary in this regard. Sheep wool had the highest air capacity with about 70%, while peat had the lowest air capacities with 18%. The peat slabs were pressed to reduce the volume for the transport, the expansion of the slabs needs time. Because the low water capacity of sheep wool at the beginning, a water reserve is missing in case of low water availability. It seems that sheep wool and perlite requires a higher and more stable supply with nutrient solution than the other substrates. The physical properties after second use of the substrates

Substrate	Before use in the experiment			After second use in the experiment		
	AC (%)	WC (%)	PV (%)	AC (%)	WC (%)	PV (%)
Sheep wool slabs	69.4 f	22.8 a	96.8 e	43.1 cd	44.1 cd	87.2 abc
Peat slabs	18.0 a	68.0 g	86.0 b	30.7 b	61.6 f	92.3 d
Coconut fiber slabs	30.6 b	52.8 e	83.9 a	20.0 a	72.3 g	92.3 d
Perlite	58.6 e	31.6 b	90.2 bcd	41.4 c	50.4 e	91.8 d
Rockwool slabs	49.2 d	41.6 c	90.7 cd	17.2 a	74.6 g	90.1 bcd

AC – air capacity, WC – water capacity, PV - volume.
Different letters indicate significant differences (Tukey P < 0.05) within one parameter.

Table 5.
Physical properties of mineral and organic substrates for cucumber cultivation [12].

were changed, in Coconut fiber and rockwool slabs the AC was below the target value.

The analyses of the mineral content of the substrates used for cucumber cultivation (**Table 6**) showed different results after the first and after second use. There was no accumulation of nutrients in the sheep wool, peat and coconut fiber slabs, but very high accumulation of NO_3 in perlite and and K in rockwool. Furthermore, In the coconut fiber slabs could be determined a high accumulation of NO_3 and Ca. In general, it can be stated that the nutrient values in the substrates do not show any unusual fluctuations. it must also be taken into account that the sorption capacity and the mineralization in the substrates are different, but the same nutrient solution was always added.

In all variants, the yield was higher in the second cultivation (**Table 7**) for all substrates tested. Furthermore, in all substrate variants the cucumber yield was

Substrate	Nutrients	First cultivation (ppm)	Second cultivation (ppm)
Sheep wool slabs	NO_3	21.6 a	23.6 a
Peat slabs		77.9 c	77.1 bc
Coconut fiber slabs		68.2 c	85.2 c
Perlite		34.7 b	65.6 b
Rockwool slabs		345.1 d	248.1 d
Sheep wool slabs	K	28.4 a	24.2 a
Peat slabs		115.7 d	72.4 c
Coconut fiber slabs		113.2 d	75.8 c
Perlite		55.3 b	56.6 b
Rockwool slabs		70.4 c	235.9 d
Sheep wool slabs	Ca	33.8 b	32.7 a
Peat slabs		64.9 d	73.7 b
Coconut fiber slabs		55.7 c	82.9 bc
Perlite		24.6 a	92.3 cd
Rockwool slabs		115.3 e	96.6 d

Different letters indicate significant differences (Tukey P < 0.05) within the cultivation.

Table 6.
Nutrient content in the tested substrates after the first and second cultivation of cucumbers.

Substrates	Yield (kg plant^{-1})			
	First cultivation		Second cultivation	
	Untreated	Treated	Untreated	Treated
Sheep wool slabs	1.08 abcd	1.94 d	8.96 e	10.07 f
Peat slabs	0.57 a	1.28 abcd	7.49 c	10.41 f
Coconut fiber slabs	0.76 ab	0.84 abc	7.51 c	8.09 d
Perlite	0.77 ab	1.16 abcd	6.68 b	8.95 e
Rockwool slabs	1.47 bcd	1.71 cd	6.11 a	9.16 e

Untreated – no application of biostimulators.
Treated – application of biostimulators (0.08% K-Humate, 0.2% Lactofol (O), spore suspension (10^7 cfu/ml) of Bacillus subtilis *FZB 24® Different letters indicate significant differences (Tuckey $_{0.05}$; comparison within one cultivation).*

Table 7.
Cucumber yield in organic substrates (sheep wool slabs, peat slabs, coconut fiber slabs) and mineral substrates (perlite, Rockwool slabs) not treated with biostimulators and treated with them [12].

higher if the variants were treated with the biostimulators. The highest yield could be obtained for cucumber cultivated in the second cultivation in sheep wool slabs and peat slabs and treated with biostimulators. The lowest cucumber yield in the second year was obtained in the substrate variants rockwool slabs and perlite.

5.2 Effects on the nutrient supply in cucumber cultivation using biostimulators

Cucumber plants were grown in substrate culture using containers (volume 8 L) filled with perlite. The perlite from Slovakia had an average dry density of 120 kg/m^3 and a grain size between 0.6 and 1.5 mm diameter, with a pore volume of 84% v/v, water capacity 45% v/v and air capacity of 39%. Investigations with the organic biostimulators were undertaken regarding the effects:

- of different humate and lactate types, whereas from *Bacillus subtilis* was used the strain FZB24®,
- of concentrations and frequencies of biostimulators applications (humates, lactates and FZB24®),
- of the best method of applications (in the rhizosphere, to the growing media or in the nutrient solution, or direct to the leaves adaxial or abaxial)
- in stress situations during growth of the cucumber plants, e.g. salt stress (EC values), suboptimal pH and suboptimal temperature
- on the nutrient uptake of N, P, K, Ca, Mg and Fe

Nutrient solution was calculated with the HYDROFER program, considering the water quality and the target value during cucumber plant growth, in order to adjust the amounts of fertilizers, salts and acids required [13]. The target values for this experiment was 170 ppm N, 50 ppm P, 260 ppm K, 150 ppm Ca, 60 ppm Mg, 3 ppm Fe, S 80 ppm, the HCO_3 content was adjusted by 90 ppm. Nutrient solution was applied using trickle irrigation 2 to 4 times a day 250 ml per irrigation cycle in period of 12–15 min.

5.2.1 Effects of biostimulators on the salt concentration (EC) in the rhizosphere

Strong fluctuation of the salt concentration (EC) can lead to an imbalance of nutrient supply in hydroponic system and can decrease plant growth and yield as it was shown for tomatoes [14]. The negative effects on cucumber plant growth if the EC values (EC 8 mS cm^{-1}) is very high could be positively influenced by application of humates and *B. subtilis* (FZB24®) separately or combined. This is probably an effect by the encouraging of root growth. Lactate (LACTOFOL) application had no effect in this regard. Even if the nutrient solution has the appropriate salt concentration (EC), in substrate culture, with increasing duration of the cucumber cultivation there could be an accumulation of salts mainly based on those nutrients, which are not necessary in the amount as applicate. [15]. This could lead to salt stress and reduced yield in crops like cucumber. Application of Humate and/or *Bacillus subtilis* FZB24® reduced this salt accumulation (**Figure 2**).

K-Humate showed the highest efficiency for EC stabilization this Humate was even more effective than *B. subtilis* (FZB24®). The mixture of all three compounds, however, was as effective as Humate alone and stabilized the salt concentration (EC) at about a value of 2 EC. The stabilizing effect of the salt concentration could be maintained over weeks after the last application indicating that there could be a

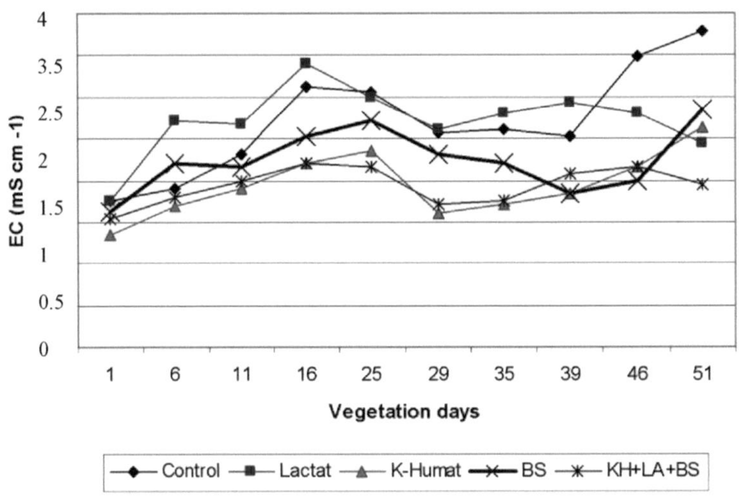

Figure 2.
Effect of biostimulators (0.1% lactate, 0.01% K-Humate, 0.2% B. subtilis *FZB24®) on EC development in the substrate during cultivation of cucumber [15].*

Variants	Shoot weight (g $plant^{-1}$)	Leaf weight (g $plant^{-1}$)	Fruits per plant
Control	676.00 d	172.33 c	7.0 ns
0.1% Lactate	817.50 bc	203.00 b	10.0 ns
0.01% K-Humate	776.33 c	203.67 b	7.8 ns
0.2% B.s FZB24®	839.33 b	202.83 b	10.4 ns
K-Humate, Lactate, B.s FZB24®	911.83 a	235.33 a	7.4 ns

Different letters indicate significant differences, LSD, $p \leq 0.05$); ns = non-significant.

Table 8.
Growth parameters of cucumber plants treated or non-treated with biostimulators [15].

sustainable culture of the bacteria and adsorption of the Humate on the perlite. The increase of EC from 1.5 to 3.6 was not so strong, but plants treated with biostimulators grew more vigorously and had in tendency a higher yield (**Table 8**).

5.2.2 Effects on pH values in the rhizosphere

For plants, which are adopted on lower pH, like tomatoes and beans [14, 16] it is well known that, pH values higher than 5.7 in substrates can disturb plant growth. In some growing media as rockwool or perlite, the initial pH is higher than 6.5. To change the pH by additional preparation of the substrates before using with a nutrient solution having lower pH takes time and is sometimes difficult. During the cultivation of the plants to change the pH value is also not always successful. Therefore, different treatments with biostimulators were tested. Application of Lactate stimulated root growth and shoot development even at pH 7.5. In general, pH values affect the nutrient availability and uptake, in particular of micronutrients. As recorded in experiments with cucumber [15], the pH of substrates in soilless culture systems, changed with the duration of cultivation (**Figure 3**) and declined in the control to 5.2. The pH of the substrates treated with biostimulators was more stable, especially if *B. subtilis* FZB24® was added to the nutrient solution as single component or in combination with the other compounds. The nutrient uptake was positive influenced even the pH was higher than recommend.

5.2.3 Effects on growth of cucumber plants

The evaluation of growth parameters of cucumber plants showed if they were treated with biostimulators then shoot and total leaf weight was significant higher than non-treated plants especially if the plants treated with all three components (**Table 8**). For the mean number of fruits harvested per plant in this short-term experiment, no significant differences between the variants, control and different treatments with biostimulators.

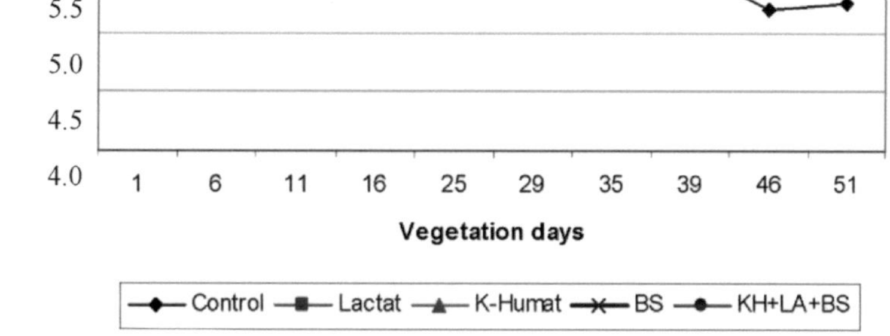

Figure 3.
Effect of biostimulators (0.1% lactate, 0.01% K-Humate, 0.2% B. subtilis *FZB24®) on pH development in the substrate during cultivation of cucumber [15].*

It can be assumed that the EC and pH stabilizing effects of the biostimulators contributed that better development of the plant, in particular if all three components are used together.

5.3 Effect of biostimulators and their application method on growing of cucumber plants

The bio-substances and Bacillus subtilis used as biostimulators had beneficial effects on plant growth in several experiments, also in stress situations, as inappropriate EC and pH value. The biostimulators were used in single applications or as a mix in the rhizosphere, but the lactate was used at the beginning as foliar-fertilizer [14]. Therefore, it came to the thought to use the biostimulators to the root zone and on the leaves. The aim was to investigated which treatment is the most effective one.

Experimental design is shown in **Table 9**, eight different treatments were compared with the control.

K-Humate (Fa. Humintech), Lactate (Fa. ECOFOL, **Table 2**) and *B. subtilis* FZB24® (Fa. ABITEP GmbH) was applied on leaves or on the substrate used in this experiment. Quantity and concentration of applied substances were deduces from previous experience [14].

Cucumber plants were planted and cultivated in containers (volume of 7–8 liters) filled with Perlite. Nutrient solution was calculated following the HYDROFER program [13] to adjust the amounts of fertilizers, salts and acids required according the values (170 ppm N, 50 ppm P, 260 ppm K, 150 ppm Ca, 60 ppm Mg, 3 ppm Fe, 90 ppm HCO3). By trickle irrigation 2 to 4 times a day 250 ml per irrigation cycle was applied in period of 12–15 min.

Additional the plants were treated with one of the treatments, 20 ml per container and plant, three times in weekly intervals in following development stages: the first treatment in 5–6 leafs stage; second in 7–8 leafs stage; third in 9–10 leafs stage. For the variant -Leaf application- the different treatments were sprayed on the surface of leafs. In case of variant -Root application- the different treatments were given to the substrate and thereby into the rhizosphere of the plants, in the same amount and frequency.

5.3.1 Shoot development

The application of biostimulators three times in the growing stage (week 4, 5, and 6) affected development and yield of cucumber plants. The application of the biostimulators stimulated the growth represented by a higher fresh matter of shoots and leaves in most variants (**Figure 4**). Obviously, the location of application was important for the effect of the biostimulators. The application to the root zone

Treatment	Concentration of substances	Leaf application	Root application
Control	—		
Lactate	0,08%	X	X
K-Humate	0,2%	X	X
Bacillus subtilis (FZB24)	Spore suspension (10/cfu/ml)	X	X
Lactate +K-Humate + *Bacillus subtilis*	Above mentioned concentration	X	X

Table 9.
Concentrations and application patterns of biostimulators used in the experiment [17].

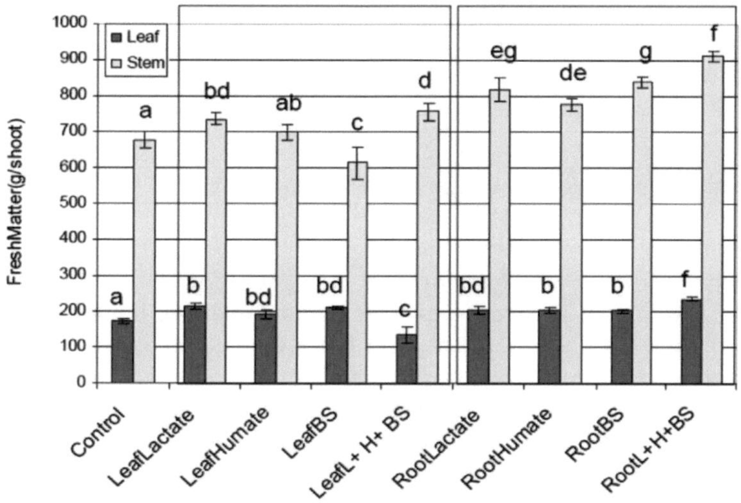

Figure 4.
Fresh matter of leaves and shoots of cucumber plants after biostimulator application (lactate, K-Humate, B.subtilis BS) on leaves and roots respectively [17]. Different letters indicate significant differences (LSD, P = 0.05).

induced in each case to a higher fresh matter compared to the control. If the Biostimulators were applied on the leaves, the effect on shoot fresh matter was not as strong in comparison to the application in the root zone. When *B. subtilis* (FZB24®) was used, the fresh mass of the shoot was even lower. It can be stated, if the combined biostimulator with all substances was applied the effect was different if applied over the roots was a stimulating effect visible, whereas if applied over the leaves an inhibiting effect for cucumber plant growth.

Comparing the effect of the treatments, also the quality of shoots and leaves seems to be different and effects on the weakness against fungi's could be expected. This effect was also found in experiments with Water spinach [18] however, in these experiments the effect on the root growth was much stronger than on the shoot growth.

Comparing the ratio between shoot and leaf fresh matter (**Figure 5**) there are no significant differences between all variants, but it seems the leaf application stronger stimulateed the leaf growth than shoot growth resulting in a lower ratio.

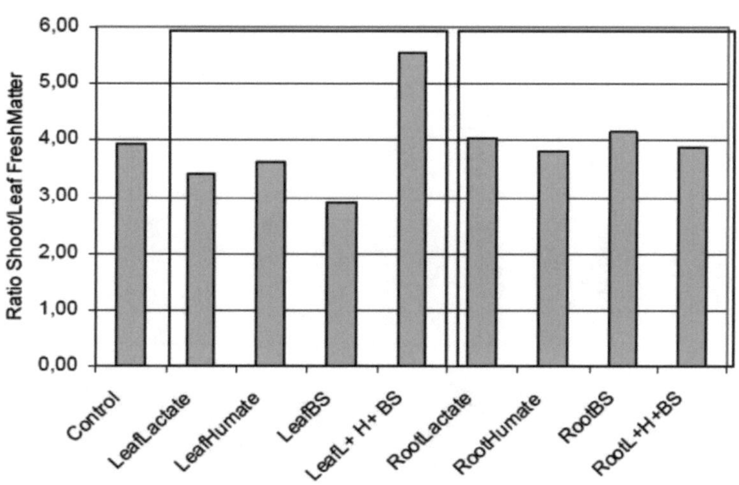

Figure 5.
Effect of application biostimulators (lactate, K-Humate, FZB24®) on leaves and roots respectively on the ratio of shoots and leaf biomass after finishing the experiment [17]. No significant differences.

However, the leaf treatment with the combination of all substances led to a reduction in leaf growth. After application of biostimulators via roots more or less the same ratio was found as in the control indicating the shoot and leaf growth was stimulated in the same manner.

5.3.2 Quantity of fruit harvest and marketable fruit quality

The number of fruits harvested from the eight variants and the control was higher on average when the biostimulators were applied to the rhizosphere directly at the roots (**Figure 6**). The fresh weight of all cucumbers with market quality was about 500 g, therefore the number of fruits is representative for the cucumber yield. In particular, the treatment with biostimulators on the substrate that means to the root system, resulted in a higher yield at the first harvest. The number of fruits finally harvested was considerably higher after treatment of roots with Lactate and *B. subtilis* (FZB24®). The number of marketable fruits was higher than in the control in most variants treated with biostimulators.

For future experiments should include more applications also during the fruit set because these additional applications could enhance the yield further, this could be especially important in long time cultivation.

The different application methods with Biostimulators on the leafs and to the roots have not only an influence on cucumber plant growth and yield, but also on the amount of marketable and non-marketable fruits (**Figure 7**). The percentage of non-marketable fruits (C class) was more than 25% in the control and could be reduced by leaf application of each Biostimulator investigated until 20% and even until 10% if substances were applied over the roots.

5.4 Use of biostimulators to reduce abiotic stress in cucumber plants

Different bacteria and in particular *Bacillus subtilis* are well known for their effects against soil-born fungal and bacterial diseases. There are selected strains in this regard with good effects in the field of plant protection. The strain *Bacillus subtilis* FZB24 is in addition capable of evolving different kinds of stress protective

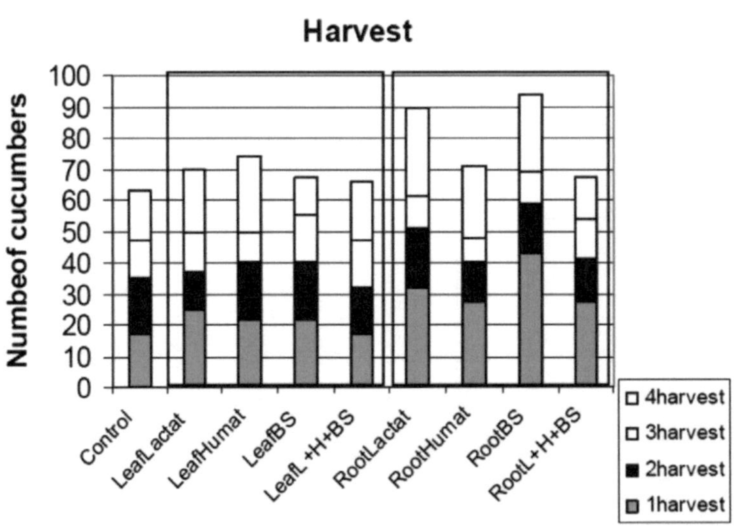

Figure 6.
Effect of application biostimulators (lactate, K-Humate, FZB24®) on leaves and roots respectively on number of marketable fruits in four harvesting periods of 9 days each. No significant differences.

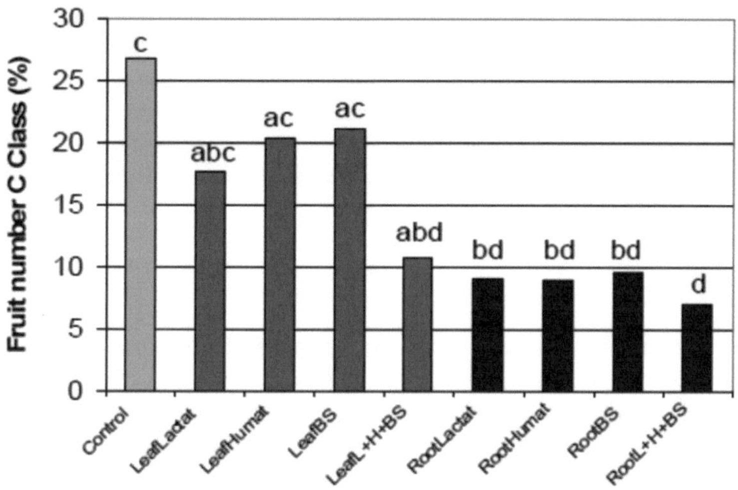

Figure 7.
Effect of application biostimulators (lactate, K-Humate, FZB24®) on leaves and roots respectively on the percentage on non-marketable fruits. Different letters indicate significant differences (chi-square-test, P = 0.05).

mechanisms including stimulation of plants' self-defense mechanisms, as it was demonstrated in chapter 5.2.1 and 5.2.2 in case of suboptimal pH, EC values. Furthermore, *Bacillus subtilis* can accelerate plant growth and stimulate the process of formation of plant organs. Furthermore, *B. subtilis* can increase the unspecific resistances of plants against stress conditions, such as extreme high temperatures, frost, drought, strong radiation, and deficiency of plant nutrients [18, 19].

Application of lactates in the form of LACTOFOL (**Table 2**) tends to reduce plant stress under suboptimal pH levels of nutrient solution (chapter 5.2.2) [6, 14, 16]. Introduction of Lactate as LACTOFOL into the growing system increases availability of micro- and macro-nutrients for plants. Investigations have shown that lactates have more stable bonds with several metal ions than other chelates do. Therefore, lactates have been used as foliar fertilizers and as well as bioregulators.

In greenhouse cultivation of cucumbers in soilless culture systems, two stress situations are of importance, the pH and the temperature stress. In an experiment was the aim to investigate the physiological effect of a biostimulating complex consisting of *B. subtilis* (FZB24®), K-Humate and Lactate as LACTOFOL (**Table 2**) on the growth of cucumber. The aim was to investigate the photosynthetic reactions to clarify first stress responses in the cucumber plants. Measuring of Chlorophyll Fluorescence is a very useful method for evaluation of plants' photosynthetic conditions and a tool in non-invasive stress detection and its subsequent evaluation [12].

5.4.1 Experimental design

(*Cucumis sativus* L.) cv. Jessica was cultivated in 'Mitscherlich' container volume 8 L filled with perlite with a physical and chemical properties as described in **Table 3**. The same nutrient solution was used as described in chapter 5.3. The cucumber plants were cultivated in a climatic chamber at 25°C and 80% RH.

One half of the cucumber plants were not treated with the Biostimulator solution. The substrate of the other half of the plants was treated once a week with 300 ml of biostimulator solution (*B. subtilis* (0.2%) + K-humate (0.01%) + lactate (0.1%) per container. Treatments coincided with the following plant developmental stages, first treatment at 5–6 leaf stage (week 1); second at 7–8 leaf stage (2 weeks) and the third at 9–10 leaf stage (3 weeks).

After the last treatment with the biostimulator solution (4 weeks), the stress factor was applied. For the pH stress experiment, pH values were adjusted to a suboptimal level (pH 3.2) by adding H_3PO_4 to the nutrient solution. This pH stress was maintained for 1 week. For temperature stress, temperature in the growth chamber was lowered from 25 to 6°C for 3 h.

5.4.2 pH stress for cucumber plants

One week after transplanting the cucumber plants the chlorophyll fluorescence Fv/Fm-value increased from 0.760 (**Figure 8**) to 0.790 in plants treated with biostimulating complex and 0.770 in plants without treatment. A drastic decrease in electron efficiency was observed after imposition of a strong lowering the pH value. Between the 4th and 5th measurements, Fv/Fm of treated plants decreased to 0.747 and that of the non-treated ones even to 0.654.

Whereas without biostimulator treatment, the fluorescence Fv/Fm-value was slightly decreasing after second and third week. In the time of the pH stress influence in the fifths week, it is visible, the stress effect was much stronger (**Figure 8**), where the lowest Fv/Fm value was 0.620. Four weeks later the plants treated with the biostimulator treatment recovered completely from the stress, but the plants without Biostimulator were still in weak condition. The plants treated with biostimulator mixture showed a higher electron efficiency of photosystem II (0.765 Fv/Fm-value) at the end of the experiment as compared to the plants without treatment (0.670 Fv/Fm-value).

The final evaluation of cucumber plants showed that this plants treated with biostimulators had significant shorter shoot length and but heavier weight than non-treated plants (**Table 10**). The roots were longer than non-treated plants and treated plants also yielded some marketable fruits (data not shown). Obviously, the biostimulators mixture was effective for reduction of the pH-stress.

5.4.3 Temperature stress for cucumber plants

Temperature stress can be happens in greenhouses if the heating system is not working or the ventilation is not proper functioning. Therefore, this stress was

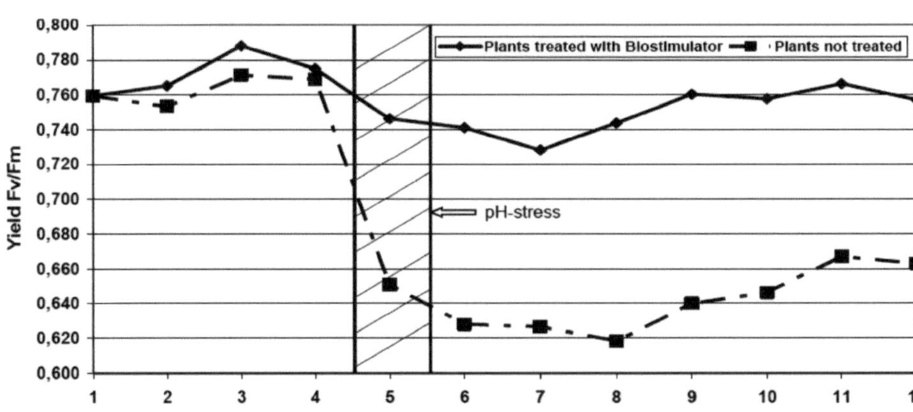

Figure 8.
Electron efficiency in photosystem II of cucumber plants treated with biostimulators mixture before and after pH-stress (pH 3.2) for one week. [12].

Variants	Shoot length (cm)	Shoot weight (g/ plant)	Leaf weight (g/plant)	Leaf area (cm^2/plant)	Root weight (g/plant)	Root length (m/plant)
Not-treated	325.75 b	230,25 b	460 ns	7002.3 ns	275.75 ns	45.08 ns
Treated with biostimulator	315.25 a	350 a	477.75 ns	7105.2 ns	265.5 ns	52.93 ns

Different letters indicate significant differences (LSD, p = 0.05); and ns = non-significant).

Table 10.
Growth parameters of cucumber plants treated with biostimulators mixture prior to pH- stress (pH 3.2) for one week [12].

given for a short time from three hours and the lowest temperature was 6°C. This stress was applied by lowering the air temperature in the climate chamber right after the third treatment with the biostimulating mixture.

The Fv/Fm parameter had the same pattern as in case of pH stress. However, Fv/ Fm development showed its peculiarities (**Figure 9**). Measurement 1 to 4 gave equal electron efficiency levels. After temperature stress, Fv/Fm values decreased considerably indicating a reduction in photosystem II efficiency. Only the treated plants were able to reach higher levels of Fv/Fm after stress and could recover much better than non-treated plants.

At the end of this temperature stress experiment, the plant growth parameters were also determined (**Table 11**). The effect of the biostimulator mixture led to a

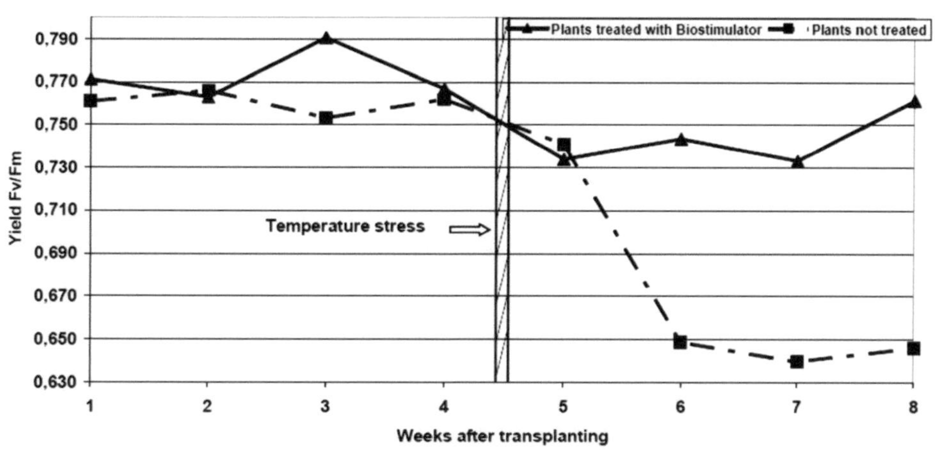

Figure 9.
Electron efficiency in photosystem II of cucumber plants treated with biostimulators mixture before and after low temperature treatment with 6°C for 3 h. [12].

Variants	Shoot length (cm)	Shoot weight (g/plant)	Leaf weight (g/plant)	Leaf area (cm^2/plant)	Root weight (g/plant)	Root length (m/plant)
Not-treated	341,75 b	244,5 b	477,75 ns	7333,7 b	206 b	4,86 b
Treated with biostimulator	390 a	358,5 a	548,25 ns	9637,4 a	321,25 a	6,98 a

Different letters indicate significant differences (LSD, p = 0.05); and ns = non-significant.

Table 11.
Growth parameters of cucumber plants treated with biostimulator mix prior to temperature - stress at 6°C for 3 h [12].

significant difference in all parameters as compared to the non-treated plants, except for the leaf area.

The question is what could be the reason, that the temperature stress was less disturbing the cucumber plant if the plants were several time treated with Biostimulator. The biostimulators used in this experiment had shown also in previous experiments a positive reaction on the root growth [14]. It can be assumed plants with well-developed root systems have higher resistance against different stress situations. Therefore, a correlation between the green biomass of the cucumber plants and the root mass were calculated. In the experiments without treatments with biostimulators, no correlation could be found (**Figure 10**).

On the other hand, those cucumber plants treated with biostimulators showed a very close correlation (R^2 linear =0.949) between green biomass and mass of roots (**Figure 11**). This close relationship confirms the hypothesis that increases in root mass lead to formation of larger shoots and leaf mass even under stress conditions if treated with some biostimulators.

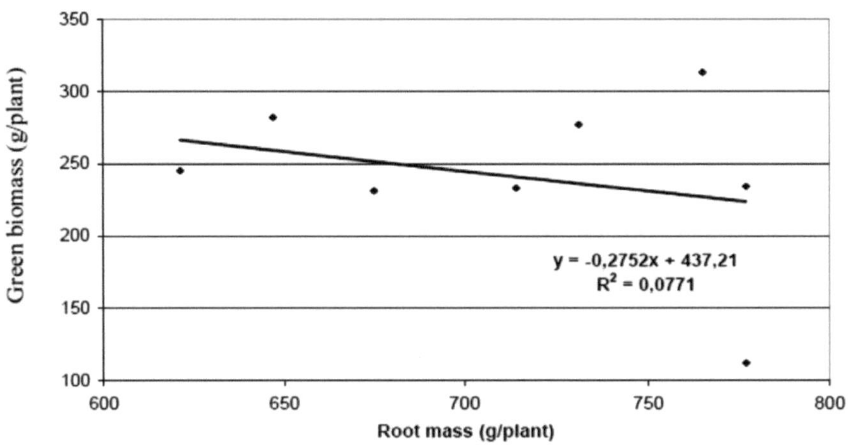

Figure 10.
Correlation between green biomass and root mass of cucumbers in plants exposed to pH and temperature stress condition, without biostimulators treatment [12].

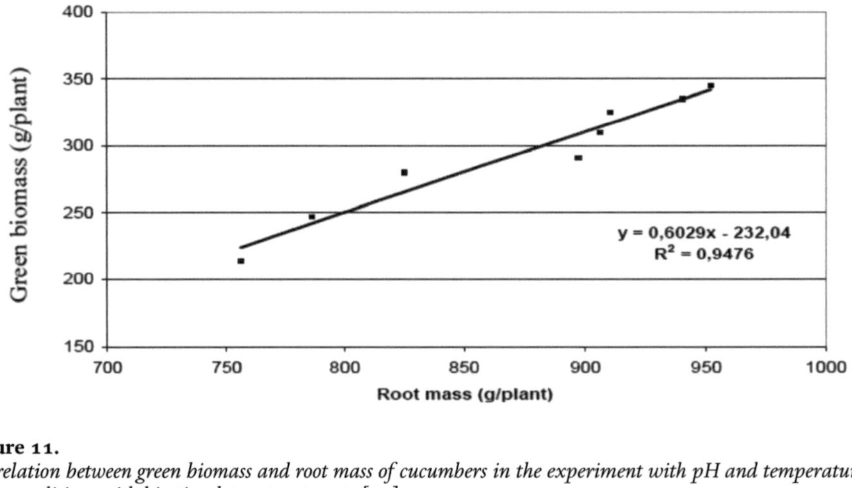

Figure 11.
Correlation between green biomass and root mass of cucumbers in the experiment with pH and temperature stress condition, with biostimulators treatments [12].

6. Conclusions

In general can be stated, intensively in greenhouses cultivated crops such as cucumbers, suffer often from inadequate abiotic or biotic growth conditions in particular in the rhizosphere. Many studies were done to find growth promoters or biostimulators to stabilize the production process or to enhance plant growth of cucumbers under these conditions. One task was to find substrates with appropriate physical and chemical properties for root growth of cucumbers desirable from organic materials. Sheep wool slabs, peat slabs and coconut fiber slabs were therefore successful tested with and without Biostimulator treatments and compared with perlite and rockwool slabs. It was to decide which of the biostimulators and their modifications should be tested and could be recommended. Based on the research with different Humates the K-Humate was selected. From the different lactates the LACTOFOL"O" (CO. ECOFOL) a foliar fertilizer was chosen after many previous experiments. From the different strains of Bacillus subtilis the strain FZB24® was successful tested and can be recommended. Under the growing conditions in soilless culture, using different substrates a buffer is missing completely or partly for regulation of the nutrient availability and regulation of the sorption capacity as in the natural soil. Therefore, different stress situations can occur, the Biostimulators used in all investigations as single component or in treatments combined in order to reduce such stress situation. Very successful was used for cucumber plants growing in substrate with a high EC value the application of K-Humate and *B. subtilis* (FZB24®) as single component and combined the salt concentration could be stabilized convenient for cucumbers. The mixture of all three components, however, was as effective as Humate alone and stabilized the salt concentration (EC) at about a value of 2 EC.

In general, pH values affect the nutrient availability and uptake, in particular of micronutrients. The pH of substrates in soilless culture systems, changed with the duration of cultivation. Application of Lactate (LACTOFOL) and *B. subtilis* (FZB24®) stimulated root growth and shoot development even at pH 7.5 and the pH of the substrates treated with biostimulators was more stable.

Application of the biostimulators solution with all three components BS-FZB24® (0.2%), K-Humate (0.01%), and LACTOFOL "O" (0.1%) were tested regarding their effects in case of strong but short time pH and temperature stress the growth of cucumber plants. The chlorophyll fluorescence Fv/Fm value showed a positive effect of the curative biostimulator treatments under the stress counteraction in plants. Results showed that there was strong correlation between green biomass of treated cucumber plants and their root mass. It can be assumed that the effect of stress prevention by the biostimulator was based mainly on enhancing the root growth.

In the experiments using the biostimulator for stabilization of cucumber plant growth, the biostimulators were applicate only in the rhizosphere that means direct to the roots. Lactate was originally developed and used as foliar-fertilizer. Therefore, it came to the thought to use the biostimulators to the root zone and on the leaves. Following the experimental results, it can be assumed, the application of the combined biostimulators with all substances if applied over the roots was a stimulating effect visible, whereas if applied over the leaves an inhibiting effect for cucumber plant growth. The number of fruits harvested from all variants and the control was higher on average when the biostimulators were applied to the rhizosphere directly at the roots in comparison to the leaf application.

Author details

Michael Henry Böhme
Department Horticultural Plant Systems, Faculty of Life Sciences, Humboldt University of Berlin, Berlin, Germany

*Address all correspondence to: michael.boehme@hu-berlin.de

References

[1] Geissler T. Böhme, M. Lankow (1991). Gemüseproduktion unter Glas und Plasten -Produktionsverfahren - Deutscher Landwirtschaftsverlag Berlin, Hochschullehrbuch; 6 Auflage.

[2] Schippers, B. 1992. Prospects for management of natural suppressiveness to control soilborne pathogenes. In E. C. Tjamos, G. C. Papavizas, & R. J. Cook (Eds.), Biological control of plant diseases, progress and challenges for the future (NATO ASI Series A: Life Sciences), Vol. 230, pp. 21–34). New York: Plenum Press.

[3] Kavetskiy, S.; M. Kraynyuk; G. Hofman; M. Böhme and O. Cleemput (2003). Soil Quality and fertilization. Tempus Tacis, NAUU, Kiev. ISBN 966-8006-20-8, p. 103–167

[4] Böhme, M., M. Dewenter and A. Gohlke (2020). Aquaponics using Asian leafy vegetables – potential and challenge. Acta Hortic. 1273. ISHS 2020. DOI 10.17660/ActaHortic1273 p. 155-122

[5] Pavlova, A., Batschvarov, P. 1992. Listno Podchranvane na Rastenijata cac Suspensioni torove LACTOFOL. Priloshenie na Suspensionite Torove, Proceedings: 1-28, Sofia

[6] Shaban, N., Manolov, I., Khadum, E., Rankov, V. (1995). Complex Assesment of the Effect of Suspension Foliar Fertilizer LACTOFOL 'O' on Cucumber. Agroeco, Proceedings, Plovdiv, 2: 33-41

[7] Frimmel F.H. and R.F. Christman (1988). Humin Substances and Their Role in the Environment. 1st ed. Vol.1., Wiley-Interscience Publication: New York.

[8] Krebs, B., Höding, B., Kübart, S., Alemayehu, M. W., Junge, H., Schmiedeknecht, G., Grosch, R., Bochow, H., & Hevesi, M. (1998). Use of Bacillus subtilis as biocontrol agent. I. Activities and characterization of Bacillus subtilis strains. Zeitschrift fur Pflanzenkrankheiten und Pflanzenschutz, 105, 181–197.

[9] Elsorra, E., Idris, H., Bochow, H., Ross, H., & Borriss, R. (2004). Use of Bacillus subtilis as bio- control agent. VI. Phytohormonelike action of culture filtrates prepared from plant growth promoting Bacillus amyloliquefaciens FZB24, FZB42, FZB45 and Bacillus subtilis FZB37. Journal of Plant Diseases and Protection, 111(6), 583–597, ISSN 0340-8159.

[10] Junge, H., Krebs, B., & Kilian, M. (2000). Strain selection, production, and formulation of the biological plant vitality enhancing agent FZB24® Bacillus subtilis. Pflanzenschutz-Nachrichten Bayer, 1(1), 94–104.

[11] Kilian, M., Steiner, U., Krebs, B., Junge, H., Schmiedeknecht, G., & Hain, R. (2000). FZB24® Bacillus subtilis – mode of action of a microbial agent enhancing plant vitality. Pflanzenschutz- Nachrichten Bayer, 1 (1), 72–93.

[12] Böhme, M., J. Schevchenko, I. Pinker , S. Herfort (2008). Cucumber grown in sheepwool slabs treated with biostimulator compared to other organic and mineral substrates. Acta Hortic. 779, p. 299–306,

[13] Böhme, M. (1993). Parameters for calculating nutrient solution for hydroponics. Eighth international congress on soilless culture, Hunters Rest, Proceedings, Wageningen, 85–96.

[14] Böhme, M. (1999). Effects of Lactate, Humate and *Bacillus subtilis* on the growth of tomato plants in hydroponic systems, Symposium on growing media & hydroponics, Windsor, Canada, Acta Hortic. (ISHS) 465, p. 231-239

[15] Böhme, M., I. Pinker, Y. Schevchenko (2007). Effects of biostimulators on nutrient supply in vegetables.: Procc Gent.. pp. 6:

[16] Böhme, M., Shaban, N. and Abdelaziz, O. (2000). Reaction of Some Vegetable Crops to treatments with Lactat as Bioregulator and Fertilizer. Acta Hortic. 514, 33–40.

[17] Böhme, M., Schevtschenko, J. and Pinker, I. (2005). Effect of biostimulators on growth of vegetables in hydroponical systems. Acta Hortic. 697:337–344.

[18] Hoang, T. L. (2003). Untersuchungen zur Wirkung von Huminsäure auf das Wachstum und die Nährstoffaufnahme von Tomaten (*Lycopersicon esculentum* MILL) und Wasserspinat (*Ipomoea aquatica* NORSSK) [PhD thesis], Humboldt-Universität zu Berlin, Landwirtschaftlich-Gärtnerische Fakultät.

[19] Hoang, T. L., & Böhme, M. (2001). Influence of humic acid on the growth of tomato in hydroponic systems. Acta Horticulturae, 548, 451–458.

Chapter 7

Soil and Agronomic Management for Cucumber Production in Nigeria

Bernard Ndubuisi Okafor and Japhet J. Yaduma

Abstract

Cucumber (*Cucumis sativus. L*) is an important crop; widely cultivated in different agroecologies of Nigeria. Its production continues to gain attention in Nigerian communities because of their nutritional and economic values. Average yield per/ha is below world average. Factors responsible for the low yield include inappropriate farming systems, climate change, pests and diseases infestation, poor access to credit facilities, inappropriate method of cultivation, distance to market and low availability of land. Important but often neglected is the quality/fertility status of cucumber producing soils of Nigeria. Fertiliser use is inadequate and application is often based on blanket recommendation. Low soil quality and poor agronomic management have contributed to >40% decrease in yield. With good agricultural practices and soil management, optimum yield can be attained.

Keywords: cucumber, soil, agronomy, Nigeria

1. Introduction

Demand for vitamins and minerals are highly dependent on vegetables. Cucumber and other fruit and leafy vegetables are in high demand because of their nutritional and economic values. According to [1], cucumber production has the capacity to enhance agricultural production, economic empowerment and food security. They are consumed fresh, as desserts in after meals, juice or in combination with other food materials. Cucumber production in Nigeria is majorly for local consumption, although Nigerian cucumbers are sometimes exported to neighbouring countries of West Africa like Chad, Cameroun, Niger and Benin Republics. Due to its importance, it ranks among major horticultural crops cultivated in Nigeria. Others are citrus, mango, African star apple, watermelon, banana, avocado pear and pineapple rank among major crops in Nigeria. Like most vegetables, its production is profitable due to high amount of cash income per unit area compared to some other crops. Cucumber production in Nigeria is usually under small scale production. Although commercial (large scale) production is also practiced under plantation farming. Some factors limiting the productivity of Nigerian soils for cucumber production include low fertility, slope, poor effective depth, stoniness/high gravel content and low nutrient/moisture retention. Good agronomic practices such as regular weeding, timely irrigation, fertiliser application and prompt harvesting are necessary for the attainment of high yield and production of quality fruits.

1.1 Factors affecting cucumber production in Nigeria

Several factors affect the attainment of sufficiency in cucumber production in Nigeria. These factors as highlighted by [2–5].

a. Climate change

b. Scarcity of improved seeds

c. Improper or inadequate fertiliser application

d. Inappropriate spacing

e. Pest and disease attacks,

f. Poor storage facilities

g. Distance to market

h. High cost of labour/capital

i. Farm size

j. Volume of agrochemicals used

k. Availability of land

1.2 Varieties of cucumber cultivated in Nigeria

According to [6, 7] there are many varieties cultivated in Nigeria. However, it should be noted that not all varieties bear the same quantity of fruit. Farmers obtain hybrid seeds from seed companies or extract seed from previous planting seasons for replanting. This also affects the eventual yield. Available cucumber varieties in Nigeria include

a. Market more, b) Poinsett, 3) Marketer 4) Ex rantan, 5) Ashley, 7) Royal, 8) Belt alpha, 9) Regal 10) Unbeit. Poinset has high yielding potential among the cucumber varieties cultivated in Nigeria [8].

2. Cucumber production in Nigeria

Cucumber is cultivated in every part of Nigerian agro ecology. This cuts across the rainforest to the savannah zones of Nigeria with the production pattern and volume varying from place to place. The 5 highest cucumber producing states in Nigeria are Plateau, Kaduna, Katsina, Kano and Benue. Others with high production capacity are Enugu, Ebonyi, Akwa Ibom, Oyo, Cross River, Rivers and Nassarawa.

The southern part of Nigeria enjoys between 6-7 months of rainfall, with an average of 1500 mm in the rainforest Zone and > 2000 mm in the Niger Delta region (Bayelsa, Delta, Rivers, Cross River and Akwa Ibom states). In Nigeria, all agro ecological zones support cucumber production but tree crops farming is more commonly practiced in the south due to high rainfall. Many exotic vegetables are produced majorly in the Guinea and Sudan savannah agro ecological zones of

Nigeria, which enjoy relatively low humidity and discourage growth of pest and diseases. Since the southern part of Nigeria enjoys bi- modal pattern of rainfall and a characteristic dry season between October/November of one year and February/March of the successive year, planting of cucumber can be practised all year round being a short duration crop of 45–55 days. Supplemental irrigation may be needed in the both south and northern parts of Nigeria are drought remains an impediment to cucumber production while excessive rainfall encourages build-up of pest and disease. States with high production capacity in Nigeria include Plateau, Kaduna, Katsina, Kano, Benue, Enugu and Ebonyi among others.

3. Production/agronomy

3.1 Climatic requirements

Cucumber requires a warm climate. Optimum day and night temperatures are 30 °C and 18-21 °C respectively. It is relatively vigorous and stand establishment is not a serious problem provided appropriate land preparation is carried out. Furthermore, soil temperature, fertility and moisture must be adequate. A well-drained soil, sandy loam to sandy clay loam is an advantage for the plants to achieve excellent establishment.

3.2 Field establishment and management

Propagation of cucumber could be through direct seeding or transplanting after nursery operations. Transplanting is best done when the ground is still wet enough to support the seedlings after transplant. Cucumber can be grown as monocrops or as intercrops with other arables or with citrus seedlings [9], *Carica papaya* [10]. In intercropping Cucumber with other crops like Pawpaw. It is important to introduce the cucumber before the time of flowering for better nutrient use efficiency by the cucumber [11]. Delayed introduction reduces vigour [12].

Staking is very necessary on the field in order to improve yield. According to [13], staked cucumber performed better than unstaked cucumber. Intrarow spacing of 50 cm -100 cm is recommended [14]. Pruning is also required as it helps to increase light penetration in the farm and reduce build-up of pest and diseases. A spacing of 50 cm × 50 cm is recommended for cucumber cultivation in Nigeria. With pruning, a yield of 571.87 kg/ha was obtained while no pruning produced a yield of 301 kg/ha as spacing of 50 cm × 50 cm gave yield of 581.59 kg/ha. At 50 cm × 100 cm yield obtained was 291.78 kg/ha while 100 cm × 100 cm spacing produced 437.04 kg/ha [15].

3.3 Pest and disease management

Pests and diseases which affect cucumber production in Nigeria include Cucumber mosaic virus, Downy mildew [16]. The variety and type of agrochemical used has strong influence on the reduction of insect pest infestation and severity [17]. Control measures include manual weeding, chemical and physical control measures. These include farm hygiene, manual eradication (depending on farm size) and use of chemicals. Weeding can be done 2–3 times before harvest. Inadequate weeding frequency affect yield significantly through yield decline [18]. It can also serve as weed control if planted as an intercrop. On the other hand, Intercropping of cucumber with Turmeric and Ginger can suppress Cucumber Mosaic Virus [19] while *Solanum torvum* and *Tithonia diversifolia* can suppress nematodes in cucumber [20]. *Hyptis suaveolens* and *Centrosema pubsecens* extracts can also be used for control of cucumber beetles [21].

3.4 Irrigation

Supplemental irrigation may be needed from time to time depending on available soil moisture. According to [22], 20% water deficit is recommended for cucumber production. Mulching is also an excellent practise to help conservation of soil water [23]. Under greenhouse conditions, 12.9 L of water is adequate [24]. Different methods used for irrigating cucumber in Nigeria include sprinkler and drip irrigation. Others are use of watering can (small scale production). In most cases, production of cucumber is under rainfed condition. Irrigation during flowering needs to be done with caution to avoid flower abortion. Irrigation in small scale cucumber farming is recommended in the early hours of the day or evening time to avoid high loss of moisture due to evapotranspiration.

4. Nigerian soils and their management for cucumber production

Nigerian soils show great variability in their properties; from the acid sands of sedimentary rock formation to the basement complex soils and coastal plain sands. pH varies from very acidic (<5 to >7). Constraints include sandiness, shallow depth, acidity (top and sub soil), and low fertility among others. Properties of Nigerian soils selected across different agro ecologies of the country are presented in **Table 1**. The soils have high base saturation but relatively low to medium effective cation exchange capacity and Total exchangeable bases.

Proper management is needed to attain sufficiency in cucumber production. Soil management is the meticulous use of soils through effective and timely combination of factors and practices which can improve soil quality and increase yield. It can be further defined as various activities carried out on the soil to improve and sustain it for optimal productivity in order to enhance food, fibre and timber production. Many agricultural practices affect cucumber producing soils. These include practices such as tillage and planting operations, irrigation, fertiliser application and use of agro chemicals for pest and disease control. Adequate soil management involves the use of soils for purposes for which they are most suited. Requirements for good Soil management includes understanding soil capability for cultivation of various crops, management of problems associated with soils, such as pH, water, fertility and temperature. Problems associated with soil management of Nigerian cucumber soils include inadequate soil survey classification and fertility assessment, lack of communication between land users and government on soil information, absence of conservation practices in farming systems,

State	pH	OC	Tot.N	Av.P	Ex. K	TEB	ECEC	BS
		g/kg		mg/kg		cmol/kg		%
Kano	5.2	9.7	2.4	5.86	0.63	17.61	18.01	99.78
Oyo	5.9	9.4	2.2	9.15	0.14	19.2	19.50	98.46
Imo	4.3	18.3	4.5	1.71	0.11	10.37	15.77	70.21
Kogi	6.1	1.53	0.27	2.84	3.96	13.69	14.25	96.07

Table 1.
Properties of some Nigerian soils.

inadequate knowledge of the environment and over exploitation of soil resources [25]. Sustainable crop production can be undermined or constrained by poor soil management as crop performance is highly related to soil quality.

4.1 Soil constraints to cucumber production and the available management options

4.1.1 Low soil fertility

Due to the inherent low fertility status of many Nigerian soils and poor availability nutrients for plant growth. Cucumber producing soils hardly have sufficient nutrients for plant needs and optimal productivity of crops. Proper soil fertility is necessary as vine length, and other growth parameters affect the eventual crop yield. Therefore, timely and appropriate application of fertiliser is required to improve yield and reduce nutrient mining and soil degradation. Organic or inorganic fertilisers are mainly used for combating soil fertility problem in cucumber production. However, there is strong advocacy for use of integrated soil fertility management (ISFM). Although fertilisers have the potentials to increase yield, there is need for caution as high fertiliser rates could also affect fruit quality [26]. Use of organo-mineral fertilisers is highly recommended as they more compatible with the nature of our soils due to their slow release pattern and environmental friendliness considering the high sand content of Nigerian Cucumber producing soils.

Examples of fertilisers used in soil fertility management in cucumber production can be broadly grouped into

a. **Organic sources:** Farmyard manure e.g. poultry manure, cow dung, compost, household waste are good sources of fertiliser for cucumber. Others are poultry manure and pig manure [27]. Application rate of 5-6 t/ha poultry manure is recommended in the Northern Guinea savannah of Nigeria [28], 35 NPK 400 g/N/ha) [29]. Odeleye et al. [30], recommended 5 t/ha in the inland valley, of SWN and 10 t/ha for upland soils. For the arid zones, 120 kg/ha Poultry Manure is adequate, while 80 kg/ha cow dung is suitable [31]. Other soil fertility improvement option include the use of cassava peel and use of organo mineral fertiliser [32], 20 t/ha PM is suitable in Kano [33].

b. **Inorganic sources:** NPK, Urea, Phosphate and potash. 50-60kgNPK/ha is recommended for the Northern guinea savannah zone [34]). According to [35], cucumber production in Nigeria requires 130 kg/haN, 95 kg/ha P_2O_5 and 200 kg/ha K_2O. It is however important to note that fertiliser application should be based on soil test results. Fertiliser should be best applied at 3–6 WAP [36].

c. **ISFM**: Integrated soil fertility management is considered most appropriate for management of cucumber producing soils of Nigeria. ISFM strategies that can be used to manage soil fertility problems in cucumber production in Nigeria include use of appropriate farming systems and planting of companion crops such as marigold to suppress soil pathogens [37] timely and adequate application of fertilisers, use of cover crops and mixed cropping to allow organic matter accumulation, use of organic manure, compost and organo mineral fertiliser, adequate irrigation to help soil processes, crop rotation. Integrated soil fertility management is the best approach to managing soils under cucumber [38].

4.1.2 Soil temperature, soil water and soil compaction

Soil water is critical to crop production just as fertiliser and other factors of production. Soil fertility depends on soil water, temperature and soil density because fertiliser sources need to decompose or dissolve depending on their source (organic or inorganic) in order to make nutrients available to plant. Similarly, microbial activities which are necessities in soil fertility economy are hampered by soil temperature and density. Strategies used to manage soil temperature, water and bulk density include.

a. **Mulching**: Mulch is any materials used to cover soil surface in order to reduce evaporation, weed infestation and action of rain and wind. Mulch materials may be natural or synthetic. Soil physical properties such as temperature, water and bulk density can be managed by using biological mulch (plant residue and fresh plant parts). Biological much tend to improve better than synthetic mulch. Other management strategies include conservation tillage, use of cover crops, application of manure, compost and organic mineral fertiliser in order to increase soil organic matter base.

4.1.3 Soil erosion

Soil erosion, which may be in form of water or wind erosion, constitutes a serious agent of soil degradation and limits the availability of nutrients and soil water. Exposure of soils to the vagaries of weather and climate, increasing population density, increase in proportion of land under cultivation, cultivating lands not suitable for cultivation has caused erosion problems. Erosion causes a reduction in soil volume, lowers crop yield, increases run off and decrease in the density of vegetation. Strategies used in combating erosion include avoidance of cultivation on sloppy lands or in cases of cultivation on sloppy lands; within land ridges should be avoided. Other strategies that can be used include terracing, contour ploughing and use of Vetiver technology [39]. Vetiver grasses can be planted in cucumber farms to help stabilise soils and reduce erosion. Other methods include proper spacing, crop rotation and use of natural mulch materials to reduce the effect of torrential rain drops on soils in Nigeria.

5. Conclusion

Cucumber production in Nigeria is an ever expanding enterprise because of their nutritional and economic uses. Poor soil management leads to decrease in production. Therefore, adoption of certain soil management strategies such as use of cover crops, conservation tillage, use of mulch and vetiver grass technology could be effective soil stabilisers. These management strategies should be adequately adopted and appropriately applied for sustainable cucumber production in Nigeria.

Author details

Bernard Ndubuisi Okafor* and Japhet J. Yaduma
National Horticultural Research Institute (NIHORT), Ibadan, Nigeria

*Address all correspondence to: kpakpando2001@yahoo.com

References

[1] Adeoye I.B and Balogun O.L (2016). Profitability and efficiency of cucumber production among smallholder farmers in Oyo sates. Journal of Agricultural science: 61(4) 387-398

[2] Ume C., Onunka C and Achike A.I (2017). A pilot production of dry season cucumber using inorganic fertilizer in the University of Nigeria, Nsukka. International Journal of Agriculture, Environment and Bioresearch 2(3): 24-42.

[3] Wilcox G.I., Offor U.S and Omojola J.T (2015). Profitability of cucumber (Cucumis sativa L) production in Tai Local Government Area of Rivers State. Journal of Advanced Studies in Agriculture, Biological and Environmental Sciences 2(3): 1-6.

[4] Umeh O.A and Orjiako F.O (2018). Limitations of cucumber (Cucumis sativa L) production for nutrition security in South Eastern Nigeria. International *Journal of Agriculture and Rural* Development 21(1): 3437-3443.

[5] Ohen S.B., Umeze G.E and Cobham M.E (2014). Determinants of market participation by cucumber farmers in Odukpani LGA of cross river state, Nigeria. Journal of Economics and sustainable developments 5(2): 188-196

[6] Akanbi w.b., Ibrahim A and Aminu D (2017). Assessment of cucumber (*Cucumis sativus* L) varieties for growth and yield in South western Nigeria. *Proceedings of 35th Annual conference of Horticultural Society of Nigeria*. pp172-178.

[7] Chinatu L.N., Onwuchekwa- C.B., Okoronkwo C.M (2017). Assessment of yield and yield component of cucumber(cucumis sativus L) in South Eastern Nigeria. International of Earth science and Agriculture 3(1): 35-44

[8] Yaduma J.J., Mudi E.I., Aliyu L., Afolayan S.O., Mani H., Yahaya R.A., Shimggu C.P., Babaji B.A., Hussaini Y., Alewadole S., Ginginyu M.B., Idris B.A., Gwamaja M.Y., Mohammed S.M., Hamisu H.S., Gani M., Hudu A.S and Idris A.U (2016). Productivity of irrigated cucumber(*Cucumis sativus* (L.J) varieties as influenced by intra row spacing at Samara and Kadawa. *Proceedings of 34th Annual conference of Horticultural Society of Nigeria.* pp 346-355.

[9] Olaniyan, A.A and Fagbayide, J.A (2006). Soil nutrients balance in sweet orange intercropped with some arable crops. Journal of Plant Nutrition 30.3: 335-350.

[10] Olubode O.O., Bello R.O and Adeyemi O.R (2014). Crop productivity and weed control efficacy of pawpaw/ cucumber mixtures supplied with organo –mineral fertilizer rates. Proceedings of 32nd annual conference of Horticultural Society of Nigeria p 82-87.

[11] Olubode O.O (2012). Component interactions and nutrient dynamics in pawpaw/cucumber mixtures in a pawpaw based cropping system. Nigerian Journal of Horticultural Science 17:33-44

[12] Ajayi A.J (2020). Effect of time of sowing okra into cucumber on growth, yield and crop mixture productivity

[13] Modupeola T.O., Ajibola A.T., Abdul-Rafiu A.M., Dixon H.G., and Akinleye O.C (2016). Effect of organic fertilizer and different staking methods on growth and yield of cucumber (*Cucumis sativus*) *Proceedings of 34th Annual conference of Horticultural Society of Nigeria.* Pp 420-427

[14] Yaduma J.J., Mudi E.I., Afolayan S.O., Idris B.A., Jibril A.,

Igwe H.C., Alewadoile S.,Shuaibu S.M., Hamisu H.S., Hudu A.H., Idris A.U., Kudi D and Hudu M (2020). Productivity of cucumber (*Cucumis sativus* L) varieties under different rates fof poultry manure at Bagauda, Kano state, Nigeria. *Proceedings of 38th annual conference of Horticultural Society of Nigeria*. Pp 628-633.

[15] Akinpelu E.A., Akinfasoye J.A and Ogunleti D.O (2011). *Proceedings of 29th annual conference of Horticultural Society of Nigeria*.Pp 169-173

[16] Chinatu L.N., Onwuchekwa- C.B., Okoronkwo C.M (2017). Assessment of yield and yield component of cucumber(cucumis sativus L) in South Eastern Nigeria. International of Earth science and Agriculture 3(1): 35-44

[17] Pitan O.O.R., Jinadu G.M., Shodunke O.K., Filani C.O and Adewole M.M (2013). Influence of varietal differences and *pyrethroid* insecticides on the population densities of cucumber ninsect pests, *Dacus spp.* And *Bactrocera invadens* (*Diptera: Tephritidae*) and fruit damage. Nigerian Journal of Horticultural Science 18: 69-77

[18] Mbah M.C., Aniekwe N.L and Atu A.O (2011). Effect of weed management on the growth and yield of cucumber (*Cucumis sativus* L) intercropped with maize (*Zea mays* L) in Abakiliki, Nigeria. *Proceedings of 29th annual conference of Horticultural Society of Nigeria.* pp 269-271.

[19] Kayode A.B., Aminu –Taiwo B.R and Ojo O.D (2020) Determination of the pathological effects of cucumber mosaic virus infection in turmeric and ginger. *Proceedings of 38th annual conference of Horticultural Society of Nigeria.* pp.634-641

[20] Izuogu N.B., Baba H.S and Okelade O.O (2017). Nematicidal potentials of two botanicals in the management of root knot nematodes (*meloidogyne incognita*) infecting two varieties of cucumber (*Cucumis sativus*). Proceedings of 35th annual conference of Horticultural Society of Nigeria. Pp 1032-1036

[21] Yussuf S.Y., Musa A.K., Hamid A.A., Babarinde S.A., Alao F.O and Oyesiji A.O (2017). Effeicacy of aqeous and ethanolic extreacts of *Hypis Suaveolens* and *Centrosema pubescens* on spotted cucumber beetle *Diabrotica undecimpunctata* (*Cleoptera:Chrysomelidae*) of cucumber. Proceedings of 35th annual conference of Horticultural Society of Nigeria 1048-1054.

[22] Emeghara U.U., Onwuegbunam D.O Oguntuyi M.J and Bala U.J (2013). Varietal response of cucumber (*Cucumis sativus*) to irrigation water deficit levels at Afaka, Kaduna. *Proceedings of 33rdh annual conference of Horticultural Society of Nigeria*. Pp 493-500

[23] Haruna I.M., Peter N.L., Ibrahim A and Umar H.S (2013). The productivity of cucumber (*Cucumis sativus*) as influenced by different mulching materials and planting spacing. *Proceedings of 33rd annual conference of Horticultural Society of Nigeria.* pp 561-569

[24] Awobona T.A., Adedapo J.O., Bolaji O.W., Ijah A.A., Ogunsanwo J.A., Zakka E.J and Ridwan A.H (2020). Evaluation off water use efficiency, growth and yield of cucumber (*Cucumis sativus* L) under greenhouse conditions. *Proceedings of 38th annual conference of Horticultural Society of Nigeria.* pp362-367

[25] Akamigbo F.O.R and Nnaji G.U (2011). Climate change and Nigerian soils: Vulnerability, impact and adaptation. Journal of Tropical Agriculture, Food, Environment and Extension 10(1): 80-90

[26] Babatola L.A (2016). Effects of levels of NPK 15:15:15 on growth and storability of Cucumber (*Cucumis sativus*).Nigerian Journal of Horticultural Science 21:39-48

[27] Adebiyi E.O., Ajayi E.O., Oladosu B.O and Folarin E.A (2017). Effects of different organic manure source on growth and yield of cucumber (*Cucumis sativus L*) in forest agro-ecological zone of Nigeria. Proceedings of 35th annual conference of Horticultural Society of Nigeria. pp 39-45

[28] Bamikole J.A., Olorukoba M.M and ishaya M (2011). Influence of organic and inorganic manure at different rates on growth and yield of cucumber (*Cucumis sativus*). *Proceedings of 29th annual conference of Horticultural Society of Nigeria,* pp 359-362

[29] Anikwe N.L (2011). Effect of some soil fertility management options on the yield and income generation potentials of cucumber (*Cucumis sativus L.*) production in Abakiliki, Nigeria. *Proceedings of 29th annual conference of Horticultural Society of Nigeria*. pp 379-383

[30] Odeleye OMO., Odeleye F.O.,Babatola J.O and Adelaja B.A (2008). Effects of different poultry manure on cucumber (*Cucumis sativum*) in Southwestern Nigeria. Proceedings of 26th annual conference of Horticultural Society of Nigeria. pp 186-189.

[31] Dantata I.J (2008). Cucumber production in semi - arid zone of Nigeria as influenced by organic sources of mineral nutrition. *Proceedings of 26th annual conference of Horticultural Society of Nigeria*. pp201-204

[32] Modupeola T.O., Babajide P,A., Akinleye O.C., Yussuf R.O (2017). Growth and yield of cucumber (*Cucumis sativus*) as influenced by different organic and inorganic fertilisers *Proceedings of 35th annual conference of Horticultural Society of Nigeria.* pp585-589.

[33] Yaduma J.J., Mudi E.I., Afolayan S.O., Idris B.A., Jibril A., Igwe H.C., Alewadoile S.,Shuaibu S.M., Hamisu H.S., Hudu A.H., Idris A.U., Kudi D and Hudu M (2020). Productivity of cucumber (*Cucumis sativus* L) varieties under different rates of poultry manure at Bagauda, Kano state, Nigeria. *Proceedings of 38th annual conference of Horticultural Society of Nigeria* 628-633

[34] Anikwe N.L (2011). Effect of some soil fertility management options on the yield and income generation potentials of cucumber (*Cucumis sativus L.*) production in Abakiliki, Nigeria. *Proceedings of 29th annual conference of Horticultural Society of Nigeria* 379-383

[35] Adebayo A,G and Akintoye H.A (2007). Growing Cucumbers in Nigeria. HORTMAG 6-7

[36] Odeleye O.M.O and Adedokun M.O (2006). Response of cucumber to time of fertilizer application. *2006 research review report of NIHORT* (90-91)

[37] Adedipe A.O (2017). Yield and post-harvest quality of cucumber as influenced by companion planting. *BSc Project, Department of Crop Science and Horticulture, Federal University, Oye* 56pp

[38] Abdulkadir S.U., Mahmoud B.A and Kashere M.A (2020). Effects of fertilizer management on growth and yield of Cucumber ((*Cucumis sativus*). Proceedings of 38th annual conference of Horticultural Society of Nigeria. Pp 79-82

[39] Babalola O (2007). Soil erosion problems in Nigeria: The use of Vetiver systems Technology. 91pp. *Ibadan University Press.*

Chapter 8

Health Beneficial Effects of Cucumber

Shampa Chakraborty and Sadhana Rayalu

Abstract

Cucumber is a healthy fruit which having lots of benefits. Cucumber regulates hydration and maintain adequate blood pressure and sugar, soothes skin, helped in digestion and reduces fat and help to weight loss. Cucumber contains a plenty of potassium, fiber, magnesium, manganese and vitamin A, C, K. Cucumber has several health beneficial activities such as antimicrobial properties, hydrating and detoxification, help in digestion and weight loss, preventing cancer and other fragile bone disease. In this book chapter we have discussed about the health beneficial activities of cucumber along with its different contents.

Keywords: cucumber, health impact

1. Introduction

Cucumbers (scientifically known as *Cucumis sativus*) belong to the same botanical family as melons and squashes. Cucumbers are served into two types. "Slicing cucumbers" for fresh consumption and "Pickling cucumbers" for eventual processing into pickles. Slicing cucumbers are usually larger with skins, while pickling cucumbers are smaller having thinner skins. Cucumbers are very low in calories but contain important vitamins, minerals and high water content. Eating cucumbers may lead to many potential health benefits, including weight loss, balanced hydration, digestive regularity and lower blood sugar levels. The present chapter deals with the different health beneficial effects of cucumber fruit as well as its contents. Fresh cucumbers are one of the fantastic fruit. Eating habits of cucumber are great as we are accustomed to eat cucumber with their seeds. This habit has an outstanding health benefits. Cucumber seeds contain a wide variety of phytonutrients, including both carotenoids and flavonoids.

2. Result and discussions

2.1 Antimicrobial activity

Khan et al. [1] reported that Cucumber have poor activity against Pseudomonas aeruginosaonly. Osuagwu et al. [2] reported that ethanolic extracts of leaf have antimicrobial property on pathogenic microorganisms for three human pathogens *S. aureus*, *P. aeruginosa* and *S. typhi* except *E. coli* pathogens. Sood et al. [3] reported antimicrobial activity of seeds of Cucumber family- members Karella, Cucumber, Tinda, Kaddu, loki. All seeds extracts are effective against Serratiamarcescens,

E. coli, Streptococcus thermophilous, and showed no inhibition against Aspergillusniger, *Candida albicans*. Mallik et al. [4] reported antifungal potentials of the ethanol extract of Cucumber against six fungus. The diameter of zone of inhibition have similar activity of the standard drug, Griseofulvin.

2.2 Phytochemical analysis

Sood et al. [5, 6] confirmed the presence of various phytochemicals like tannins, cardiac glycosides, terpenoides, carbohydrates, resins, saponins and phytosterols. Phytochemical screening by color visualization in cucumber showed the presence of flavonoid, tannin, saponin and steroid. The spectrophotometric analysis showed flavonoid 0.36% (w/w), phenol 0.40% (w/w), and, the by titrimetry found tannin 2.82%.

2.3 Nutritional composition

Urooj et al. [7] reported that among all varieties, English, Zucchini and Pranic healed cucumbers have contained highest moisture and lowest in Holenarasipur and Dotted variety. Eghtedary et al. [8] reported twenty types of cucumber (*Cucumis sativus*) from different regions. Khan [9] reported 24 genotypes of cucumber based on Randomized Complete Block Design (RBCD) method.

Cucumber, sliced, raw 1.00 cup (104.00 grams) calorie-16.

Nutrient	
Vitamin K	19%
Molybdenum	12%
Pantothenic acid	5%
Copper	4%
Phosphorus	4%
Vitamin C	4%
Biotin	3%
Vitamin B1	3%
Potassium	3%
Magnesium	3%
Manganese	3%

3. Different types of nutrients in cucumber

Cucumber contains different types of flavonoids such as apigenin, diosmetin, fisetin, luleolin, quercetin, kaempferol, luteolin, naringenin, theaflavanoside I, vicenin. It also contains different types lignans pinoresinol, lariciresinol and secoisolariciresinol and triterpenes like cucurbitacin A, cucurbitacin B, cucurbitacin C, cucurbitacin D.

3.1 Medicinal uses

Shrivastava et al. [10] reported that cucumber eating in daily basis can improve hair growth and soothes skin, reduce swelling of eye. Cucumber juice can improve the skin texture and cures skin infections, eczema. A piece of cucumber on the head for a sun stroke patient can reduce the heat of the body. Cucumber can help in weight loss and the mashed seeds with sugar before meal can reduce intestinal worms and tape worms. Boiled cucumber leaves and cumin seed roasted powders can be used for throat infections. Kashif et al. [11] reported that cucumber contained high potassium so it can help to keep normal blood pressure.

3.2 Different health beneficial activities of cucumber

Sharma et al. [12] proved that cucumber juice can acts against acidity and resist to change in pH and have good carminative and antacid potential. Cucumber has also eye soothner activities. Keeping cucumber slices on the eyes for about 10 minutes relaxes our eyes and reduces puffiness around the eyes. Patil et al. [13] investigated activity of cucumber extract in laboratory animals. They proved that the aqueous extract of cucumber selected can reduce the induced bowel disease and possessed active against ulcerative colitis. Patil et al. [14] investigated that cucumber juice can have significant wound healing effect.

3.3 Hepatoprotective activity

Heidari et al. [15] investigated the activity of cucumber juice against cumene hydroperoxide induced-oxidative stress and proved that cucumber can acts as antioxidant agent. The cucumber extracts have antioxidants and radical scavenging property. The extract help to form intracellular ROS.

3.4 Hypoglycemic and hypolipidemic activity

Sharmin et al. [16] investigated Hypoglycemic effects of cucumber and proved white pumpkin, gourd has anti hyperglycemic effects on Alloxan Induced Diabetic Rats (AIDR). The extract can lessen the high lipid profiles in AIDRs. Therefore, cucumber extracts can be useful, in the therapy of diabetes, hyperglycemia and hyperlipidemia.

3.5 Hydration and detoxification

Cucumbers contains 96% water. Therefore, it helps to fill up the daily requirement of water by the body and keep body hydrated. In summers, people tend to dehydrate easily. Consuming cucumber can make hydrate and it also acts as a coolant and give relief from the summer heat. Cucumber and mint can use to make detox water which effectively eliminates toxins from the body, improves hydration and thus results in innumerable health benefits.

3.6 Reduce blood pressure

Cucumbers are a good source of potassium, magnesium and dietary fibre. These nutrients are known to lower blood pressure, thus reducing the risk of heart diseases. Research has also proved that regular consumption of cucumber juice was helpful in reducing blood pressure, in elderly people with hypertension.

3.7 Digestion

Cucumbers act as a coolant for our stomach. The soluble fibre in cucumbers helps in slowing our digestion. Also, the high content of water in cucumber makes our stools soft, prevents constipation and keeps our bowel movements regular.

3.8 Reduces blood sugar

Cucumbers are known to reduce blood sugar levels, thus being helpful in the management and prevention of diabetes mellitus.

3.9 Weight loss

Cucumbers contain 96% of water and are low in calories. There are only 15.5 calories in 100 g of cucumber. High water and low-calorie content of cucumbers, helps in reducing weight.

3.10 Skin

Cucumbers can enhance beauty and have good effects on the skin. Application of cucumber juice on skin makes it soft and glowing. Anti-inflammatory effects of cucumber naturally lighten our skin and reduce tanning. It also reduces wrinkles and fine lines.

3.11 Reduces risk of cancer

The fibre in cucumbers protects from colorectal cancer. Also, cucurbitacin present in cucumbers possesses anti-cancer properties. Tuama et al. [17] showed cucumber extract is rich in bioactive compounds and have anticancer activity with Cell lines of (IC50) with MCF 715.6 ± 1.3 and HeLa 28.2 ± 1.

4. Effect in hair, nails and breath

Cucumbers contain silica which is excellent for hair and nail care. They help in strengthening the nails and prevent them from becoming brittle. Moreover, phytochemicals present in cucumber destroy the bacteria in our mouth that cause bad breath.

4.1 Antioxidant activity

Kumaraswamy [18] demonstrated that the yellow cucumber posses more anti-oxidant activity where as While green and white cucumbers have low antioxidant property. The antioxidant activity is due to the carotenoids, phenolic flavonoids, tannins, polyphenols and lycopene, found in it. Mallik et al. [19] carried out an study on the cytotoxic activity of cucumber extract in ethanol.

	Cucumber, sliced, raw 1.00 cup (104.00 grams) calorie-16	
Nutrient	**Amount**	**DRI/DV (%)**
Protein	0.68 g	1
Carbohydrate	3.78 g	2
Fat-Total	0.11 g	0
Dietary Fibers	0.52 g	2
Calories	15.60	1
Total sugar	1.74 g	—
Monosaccharides	1.70 g	—

	Cucumber, sliced, raw 1.00 cup (104.00 grams) calorie-16	
Nutrient	**Amount**	**DRI/DV (%)**
Fructose	0.90 g	—
Glucose	0.79 g	—
Galactose	0.00 g	—
Disaccharides	0.04 g	—
Lactose	0.00 g	—
Maltose	0.01 g	—
Sucrose	0.03 g	—
Soluble Fiber	0.06 g	—
Insoluble Fiber	0.46 g	—
Other carbohydrate	1.52 g	—
Fat		
Monosaturated fat	0.01 g	—
Polysaturated fat	0.03 g	—
Saturated fat	0.04 g	—
Trans fat	0.00 g	—
Calories from fat	1.03	—
Insoluble fiber	0.46 g	—
Other carbohydrate	1.52 g	—
Calories from saturated fat	0.35	—
Calories from trans fat	0.00	—
Cholesterol	0.00	—
Water	99.04 g	—
Vitamins (water soluble)		
Vitamin B1	0.03 mg	3
Vitamin B2	0.03 mg	2
Vitamin B3	0.10 mg	1
Vitamin B3 (Niacin equivalent)	0.19 mg	
Vitamin B6	0.04 mg	2
Vitamin B12	0.00 mcg	0
Biotin	0.94 mcg	3
Choline	6.24 mg	1
Folate	7.28 mcg	2
Folate DFE	7.28 mcg	
Folate food	7.28 mcg	
Pantothenic acid	0.27 mg	5
Vitamin C	2.91 mg	4
Fat soluble vitamins		
Vitamin A international unit	109.20 IU	—
Vitamin D	0.00 IU	0
Vitamin K	17.06 mcg	19

	Cucumber, sliced, raw 1.00 cup (104.00 grams) calorie-16	
Nutrient	**Amount**	**DRI/DV (%)**
Vitamin E	0.05 IU	
Minerals		
Boron	0.00 mg	
Calcium	16.64 mg	2
Chloride	0.00	—
Chromium	0.00	—
Copper	0.04	4
Fluoride	0.00	0
Iodine	0.00 mg	—
Iron	0.29 mg	2
Magnesium	13.52 mg	3
Manganese	0.08 mg	3
Molybdenum	5.20 mcg	12
Phosphorus	24.96 mg	4
Potassium	152.88 mg	3
selenium	0.31 mcg	1
Sodium	2.08 mg	0
Zinc	0.21 mg	2

The nutrient profiles provided in this chapter are derived from The Food Processor, Version 10.12.0, ESHA Research, Salem, Oregon, USA.

5. Conclusion

Cucumber is an excellent and unusual food. A cup of cucumber contained 16 calorie with its peel (15 without) and 4 percent of your daily potassium, 3 percent of your daily fiber and 4 percent of your daily vitamin C. Cucumbers also contained small amounts of vitamin K, vitamin C, magnesium, potassium, manganese and vitamin A among them Vitamin K is very important for bone health. Vitamin K intake reduces fracture rates, and combine work with vitamin D can increase bone density and positively affect calcium balance. Vitamin K helps in building bones and the effects seem to be important for women. Low vitamin K levels were associated with low bone density in women, but not in men. Low intakes of vitamin K were associated with an increased risk of hip fractures in middle-age women.

Acknowledgements

Authors thank the SERB (PDF/2016/002856) (Govt. of India) for financial support and Knowledge Resource Center-NEERI for checking plagiarism. (KRC No.:CSIR-NEERI/KRC/2021/Jan/EMD/1). SC acknowledges Director-NEERI for his overall support.

Author details

Shampa Chakraborty* and Sadhana Rayalu
Environmental Materials Division, National Environmental Engineering Research Institute, Nagpur, India

*Address all correspondence to: shampa132@gmail.com

References

[1] D. A. Khan, F. Hassan, H. Ullah, S. Karim, A. Baseer, M. A. Abid, M. Ubaid, S. A. Khan and G. Murtaza. Anti-bacterial Activity OfPhyllantusEmblica, CoriandrumSativum, Culinaris Medic, Lawsonia Alba And CucumisSativus. ActaPoloniaePharmaceutica n Drug Research, 70(5): 855-859, 2013.

[2] G. G. E. Osuagwu and F. O. Ejikeme. The Antimicrobial Activity of the Leaves of Some Wild Cucurbitaceae Species from South-East Nigeria. Human JournalsResearch ArticleMarch 2(4): 2015.

[3] A.Sood, P. Kaur and R. Gupta. Phytochemical Screening And Antimicrobial Assay Of Various Seeds Extract Of CucurbitaceaeFamily. International Journal of Applied Biology and Pharmaceutical Technology, 3(3): 401-409, 2012.

[4] J.Mallik., R.Akhter. Phytochemical Screening and In-vitro Evaluation of Reducing Power, Cytotoxicity and Anti-Fungal Activities of Ethanol Extracts of Cucumissativus. International Journal of Pharmaceutical & Biological Archives,3(3): 555-560, 2012.

[5] A.Sood, P. Kaur and R. Gupta. Phytochemical Screening And Anti-microbial Assay Of Various Seeds Extract Of CucurbitaceaeFamily. International Journal of Applied Biology and Pharmaceutical Technology, 3(3): 401-409, 2012

[6] Jet S. Mandey, Fenny R. Wolayan, Cherly J. Pontoh, and Bonie F. J. Sondakh, Journal of Advanced Agricultural Technologies, Vol. 6, 1, 2019

[7] T. Keerthika, C. S. Devaki, F. Suma and A. Urooj. Studies on the Nutritional and Quality Characteristics of CucumisSativusVarieties .Agricultural Science Research Journal Vol. 6(4): 79-85, 2016.

[8] M. Golabadi, P. Golkar and A.R Eghtedary. Assessment of genetic variation in cucumber (CucumissativusL.) genotypes. European Journal of Experimental Biology, 2 (5):1382-1388, 2012.

[9] Z. Khan, A. H. Shah, R. Gul, A. Majid, U. Khan, H. Ahmad. Morpho-agronomic characterization of cucumber germplasm for yield and yield associated traits.InternationalJournal of Agronomy and Agricultural Research (IJAAR)ISSN: 2223-7054 (Print) 2225-3610 (Online) 6(1), 1-6, 2015.

[10] Dr. A. Shrivastava, Dr.S. Roy. Cucurbitaceae: AEthnomedicinally Important Vegetable Family.Journal of Medicinal Plants Studies., 1(4):16-20, ISSN: 2320-3862, 2013.

[11] W.Kashif, Q. M. Kamran, M. S. Jilani. Effect of Different Nitrogen Levels on Growth and Yield of Cucumber (Cucumissativus L.). J. Agric. Res. 46(3): 259-266, 2008.

[12] S. Sharma., J.Dwivedi., S.Paliwal. Evaluation of antacid and carminative properties of Cucumissativus under simulated conditions. Scholars Research Library Der Pharmacia Lettre, 4(1): 234-239, 2012.

[13] K.Patil., A Kandhare., D Bhise. Effect of aqueous extract of Cucumissativus Linn. fruit in ulcerative colitis in laboratory animals. Asian Pacific Journal of Tropical Biomedicine:962-969, 2012.

[14] K.Patil., A.Kandhare., D.Bhise. Pharmacological evaluation of ameliorative effect of aqueous extracts of CucumissativusL.fruit formation on wound healing in Wister rats.Chronicles of young scientists. 2(4),207-213, 2012.

[15] H.Heidari, M.Kamalinejad, M.Eskandari.Hepatoprotective

activity of Cucumissativus against cumenehydroperoxide induced-oxidative stress. Research in Pharmaceutical Sciences,; 7(5): 936-939, 2012.

[16] R.Sharmin., M. Khan., M.Akhter., A.Alim., A. Islam., and M. Ahmed. Hypoglycemic and Hypolipidemic Effects of Cucumber, White Pumpkin and Ridge Gourd in Alloxan Induced Diabetic Rats. J. Sci. Res. 5 (1), 161-170.

[17] A. A. Tuama, A. A. Mohammed, Saudi Journal of Biological Sciences, 26, 600-604, 2019

[18] L. Kumaraswamy. A comparative study on antioxidant activities of three cultivars of Cucumissativas (Linn). International Journal of Research in Biotechnology and Biochemistry,2016.

[19] J. Mallik., R. Akhter. Phytochemical Screening and In-vitro Evaluation of Reducing Power, Cytotoxicity and Anti-Fungal Activities of Ethanol Extracts of Cucumissativus. International Journal of Pharmaceutical & Biological Archives; 3(3): 555-560, 2012.

Chapter 9

Weed Interference and Management in Cucumber (*Cucumis sativus* L.)

Olumide Samuel Daramola

Abstract

Weed interference is a major problem in cucumber farming, leading to 45–95% yield reduction. Weed control practices employed to avoid such losses are predominantly hand weeding and herbicides application. All the weed control methods used in cucumber farming have their own limitations. Hand weeding is tedious, time consuming and associated with high labor demands. Only few herbicides are registered for weed control in cucumber, and these herbicides does not provide season-long weed control when used alone, neither can they control the entire weed spectrum with diverse physiology, morphology and time of emergence. Therefore, to optimize yield, financial and environmental costs and benefits, integrated weed management approaches are advocated. A good tillage operation, use of competitive cultivars, appropriate plant population and row spacing, application of pre and post emergence herbicides are important in reducing weed density. The combination of these approaches provides effective weed control, and helps in environmental conservation. The world is now moving toward precision weed management techniques which involve remote sensing, modelling and use of robotics to control weeds. These technologies are the future of weed management in crop production and have a substantial role to play in modern cucumber production. Right selection of one or more of these techniques with reference to environmental, socioeconomic, and geographic conditions will provide effective weed control in cucumber. Future research should therefore be focused on delivering information for the implementation of these approaches.

Keywords: Weed competition, hand weeding, herbicides, integrated weed management

1. Introduction

Weeds are one of the most important pest that reduces crop productivity. Weeds and crop plants are very similar in their demand for carbon dioxide and nitrogen from the atmosphere, water and minerals from the soil and light from the sun for their growth and development [1]. When weeds compete for these restrictive resources, the growth and development of crops are restricted, and their yield and productivity drastically reduced. The type and density of weed population, and the duration of weed-crop interference determines the magnitude of damage and yield loss inflicted by weeds on crop plants [2, 3]. Weed interference in cucumber results in high yield reduction in the range of 45–95% in different agro climatic conditions [4–6]. Growers and agricultural experts throughout the world consistently indicate that weeds are one of the most economically important pests of cucumber. Therefore, an effective weed management strategy is recognized as a necessity for an economically feasible cucumber production [4, 6].

Currently, weed management methods employed to reduce yield losses in cucumber are predominantly hand weeding and herbicides application [7]. However, these weed management systems have a number of limitations. Hand weeding is tedious, inefficient, time consuming and associated with high labor demands [1, 7, 8]. In addition, labor for manual weeding is scarce and often too expensive [1, 2, 9]. Consequently, farmers spend a large amount of time in weeding operation. Despite the effort expended in weeding by farmers, weed still cause considerable yield losses, because most of the weeding operations are not done during the critical period of weed interference, but well after the crop have suffered irrevocable damage from weeds [4, 10]. Most weed competition in cucumber is a consequence of delayed first weeding during the early stage of crop growth [7, 11]. Moreover, the efficacy of hand weeding is often compromised by the continued wet condition characteristics of the beginning of the rainy season in many agro climatic zones. Hand weeding under conditions of irrigation or high rainfall often causes weed to re-root and re-establish, necessitating several rounds of weeding to keep cucumber weed-free and avert yield losses [5].

Herbicides are quite effective in suppressing weeds in cucumber if used properly [7]. Herbicides reduce drudgery and protect cucumber from early weed competition [4, 10]. However, only a few herbicides are registered for weed control in cucumber [12, 13]. Moreover, most available herbicides do not provide season-long weed control when used alone, and single herbicide application may not control the entire weed spectrum with diverse physiology, morphology and time of emergence [4, 6, 14]. Although herbicide use alleviates the problem of labor for weeding, incorrect use may be injurious to the crop and bring about other environmental problems [15, 16].

No-tillage, conventional tillage, stale seedbed, and mulching are other options currently utilized for weed control in cucumber [15–18]. However, these weed control methods are limited and inefficient when used as a stand-alone weed management tactics [5]. Therefore, the adoption of integrated weed management (IWM) strategy is more advantageous than relying on one form of weed control. IWM involves the reduction of weed interference through a combination of two or more methods while maintaining acceptable crop yields, environment, social and economic wellbeing [1, 19]. However, in literature, information on weed interference and management methods, especially IWM strategy for improved productivity of cucumber is very scattered and not available in the form of a single document. Therefore, this book chapter is compiled to present all the available information into one document, which will be useful to all cucumber industry stakeholders like researchers, academicians, the extension community, industrialists, and growers. This book chapter covers in detail the weed flora of cucumber, their impact on cucumber and yield losses due to weed interference, different methods of weed control and IWM management strategies in cucumber.

2. Weed flora of cucumber

Diverse weed species infest cucumber but the extent of damage inflicted on cucumber crop varies with the type of weed species involved. A complete list of weed flora in cucumber grown in different agro climatic zones around the world is

Weeds	References
Abutilon theophrasti Medicus	[20, 21]
Adconopus compressus	[5, 22]
Ageratum conyzoides	[22, 23]
Ambrosia artemisiifolia L.	[5, 20, 23]
Amaranthus hybridus L.	[5, 22–26]
Amaranthus spinosus L.	[5, 23–25]
Amaranthus palmeri S. Wats.	[4, 23–25]
Amaranthus lividus L.	[5, 24, 25]
Amaranthus retroflexus L.	[23–25]
Ambrosia artemisiifolia L.	[22, 23]
Anodacristata L.	[20, 24]
Aspilia Africana	[22]
Axonopus compressors (Sw.) P. Beauv	[22, 27]
Bidens pilosa	[22, 23, 27]
Boerhavia diffusa (Linn).	[22, 27]
Chenopodium album L.	[20, 22, 23]
Chloris pilosa Schumach	[27]
Chromoleana odorata (L.) R.M. King and Robinson	[22]
Chrysopogan aciculatus (Retz.) Trin.	[27]
Combretum hispidum Laws.	[27]
Commelina benghalensis (Burn.)	[20, 24]
Commelina diffusa L.	[22]
Commelina errecta L.	[24]
Convolvulus arvensis L.	[23]
Coronopus didymus	[27]
Croton hirtus L'Herit	[23]
Crupheacarth agenensis	[24]
Cynodon dactylon L.	[22–24]
Cyperus esculentus L.	[5, 23, 24]
Cyperus rotundus L.	[5, 20, 22, 23, 26]
Digitaria horizontalis Willd.	[24]
Echinochloa crusgalli	[23]
Euphorbia heterophylla L.	[27]
Euphorbia hirta	[27]
Euphorbia glomerifera	[27]
Eleusine indica L.	[22]
Emilia sonchifolia	[20, 24]
Emilia coceinea	[22, 24]
Eragrostis atrovirens	[22]
Galinsoga spp	[20]
Impereta cylindrical	[27]

Weeds	References
Ipomoea spp	[26]
Ipomea triloba L.	[27]
Jamaica vervain	[27]
Laportea aestuans	[22]
Mimosa diplotricha C. Wright ex Sauuville	[22]
Mimosa pudica	[22]
Paspalum conjugatum	[22]
Paspalum scrobiculatum L.	[27]
Panicum maximum Jacq.	[22]
Portulaca pilosa	[27]
Portulaca oleracea	[20, 23]
Phyllantus samarus	[22, 23]
Seteria verticillata	[23]
Spigelia anthelmia L.	[23, 27]
Sida acuta	[22]
Solanum carolinense	[24]
Solanum nigrum	[23, 24]
Sonchus oleraceus	[24]
Sorghum halepense	[23],
Starchyptophetae ayenesis	[22]
Talinum triangulare (Jacq.) Willd.	[20]
Tribulus terrestris L.	[23]
Tridax procumbens	[22]
Xanthium strumarium L.	[5, 20]

Table 1.
A list of weed flora of cucumber.

presented in **Table 1**. However, major problematic weeds in cucumber include broad-leaved weed species such as members of the families Amaranthaceae (*Amaranthus retroflexus*, *Amaranthus spinosus*, *Amaranthus hybridus, Chenopodium album*) Asteraceae (*Tridax procumbens*, *Bidens pilosa, Xanthium strumarium, Ambrosia spp*); Euphobiaceae (*Euphorbia heterophylla*, *Euphorbia hirta*); Convolvulaceae (*Ipomoea spp*, *convolvulus arvensis*); Portulacaceae (*Portulaca oleracea, Purtulaca pilosa*); Solanaceae (*Solanum carolinense Solanum nigrum*), grasses weed species of the family Poaceae (*Cynodon dactylon*, *Sorghum halepense, Echinochloa crusgalli, Seteria verticillata, Digitaria spp, Paspalum spp, Panicum maximum*) and sedges of the family Cyperaceae including *Cyperus rotundus,* and *Cyperus esculentus* [24, 27–29]. The major feature of these weed species is their widespread existence and difficulty in management. Annual broad-leaved weeds like *Amaranthus spp, Chenopodium album, Solanum nigrum, Portulaca oleracea* and *Euphorbia spp* cause serious damage to cucumber due to their rapid spread, production of many seeds, high efficiency in water use and net photosynthesis [4, 5, 24, 27]. Just 1–2 plants of *Amaranthus spp* per square yard growing with cucumber throughout the crop life cycle can reduce yield by 10%, while 5–7 plants of *Amaranthus spp* per square yard can reduce cucumber yield by 50% [5]. The occurrence of many biotypes of this weed specie and its resistance to sulfonylurea herbicides also complicates its management [25]. *Chenopodium album* is able to outgrow cucumber and compete with the crop for nutrients, light and moisture. Its rapid growth and establishment rate makes it difficult to control by cultivation [20]. *Solanum nigrum* also grows rapidly and is able to out compete with cucumber vines. It is also capable of hosting pest such as white fly [20, 25]. *Portulaca oleracea* spreads quickly due in part to its large seed it production. It also harbors pest such as caterpillar moths and spread quickly between the crop rows [20]. *Bidens pilosa* is another broad-leaved weed specie with great adaptability and one of the most difficult to control in cucumber [20]. Its main features are: the extensive formation of achenes, high water use efficiency in region of prolonged drought stress and dormancy which facilitates its viability in the soil. *Bidens pilosa* is also resistant to herbicides that inhibit the acetolactate synthase, which further makes it difficult to control in cucumber [20, 30].

Generally, annual weeds are the main problem in cucumber but perennials such as *Cyperus rotundus, Cyperus esculentus*, *Sorghum halepense Cynodon dactylon* and *convolvulus arvensis* are also difficult to control and possess considerable problems to cucumber [24, 25, 27, 29]. These weed species remain alive for more than one year in spite of producing seeds in the growing season proceeding the dry season and, therefore cause significant damage to cucumber. They are difficult to control because they have the capacity to survive adverse conditions by forming extensive underground vegetative structures such as rhizomes and stolon [31]. *Cyperus rotundus,* and *Cyperus esculentus* can reproduce sexually and asexually by rhizomes and tubers, and therefore exert significant competition for moisture, carbon dioxide, light and nutrient in addition to their allelopathic effects [31]. Competition and allelopathic effects of *Cyperus rotundus* at high density may reduce cucumber yields as much as 83% [32].

3. Effect of weed interference on cucumber

Weed interference is the detrimental effects of weed on crop resulting from their interaction with each other. Weeds are considered as the most harmful pest of crops, and their interaction with crops have considerable consequences on the economy, society, and the environment [33]. They limit crop productivity and profitability, alter the ecosystem function and hamper the sustainability of the agricultural system. Yield losses and reduced profitability due to weed interference is considered one of the major problem in cucumber production [10]. Even with advanced technologies, producers record high losses due to weed interference. According to estimates, between 45–95% potential yields of cucumber is lost due to weed interference depending on the type and density of weeds growing in the crop community, duration of weed interference, stage of crop growth at which the interference takes place and the crop variety [4, 22]. Generally, losses due to weed interference in cucumber can either be direct or indirect. Direct losses due to weed interference includes damages caused by weed's allelopathic interaction with cucumber and competition for growth resources such as nutrients, water, light and space [5, 10]. Weed interference affects cucumber production indirectly by sheltering crop pest and diseases, interfering with timeliness and efficiency of harvest, increasing harvest difficulties, reducing fruit quality and consequently increasing the cost of processing [34, 35]. Weeds are potential source for diseases and pest including powdery mildew (*Podosphaera xanthii*), gummy stem blight (*Didymella bryoniae*), fungal root rot (including *Pythium, Rhizoctonia* and *Fusarium*), thrips (Thrips palmi) which may be hosted by a variety of weeds including *Portulaca spp, Amaranthus spp*, *Gomphrena celosioides* and white fly [36–39].

Weed interference in cucumber begins during the very early stages of vegetative growth [5]. Unfortunately, cucumber is not a strong competitor at the early stage, therefore weeds out grow them during the early stage of crop growth, resulting in high yield reduction [4]. Weeds that germinates at the same time as cucumber such as annual weeds like *Amaranthus spp*, *Xanthium strumarium* and *Ambrosia artemisiifolia* grows faster and maintain canopy above and below the top of cucumber. Hence, these weeds intercept photosynthetically active radiation at the expense of the crop, resulting in reduced yield [40, 41]. Furthermore, weed-inflicted shading of cucumber flowers promotes flower abortion. Although cucumber becomes less vulnerable to weed competition after the vines run out or when they become well established, the crop may take a few more weeks to close canopy. Weeds that emerge during this period may complicate harvest by concealing fruit or hampering manual picking with prickly foliage, or entangling vines, and promote fungal diseases by limiting air circulation. *Solanum carolinense* is a host for cucumber powdery mildew fungus (*Erysiphecichor acearum*), and many common weeds such as *Amaranthus spp* and *Cyperus spp* can carry cucumber mosaic virus [42].

4. Critical period of weed control (CPWC) in cucumber

The CPWC is a period in the crop growth cycle during which weeds must be controlled to prevent quantitative and qualitative yield losses [21]. It is the period when a crop is most sensitive to weed competition and therefore the time interval when it is necessary to maintain wed-free condition to prevent an unacceptable reduction in potential yield [23]. It denote the optimum timing of weed removal to prevent potential yield loss. From practical standpoint, crop yield losses from weed interference before or after the CPWC will be of limited interest. This means that weeds that are present before or emerge after the CPWC do not cause significant yield loss [43]. Studies conducted on cucumber [5, 44] have shown that weed infestation during the CPWC imposes irreversible loss and damage on the final yield, while weed control before or after the CPWC did not improve fruit yield compared with crops kept weed-free only during the CPWC [45, 46]. Weed control recommendations in cucumber are therefore made on the basis of the CPWC because they indicate the optimum time for implementing and maintaining weed control at reduced cost [47]. Although published research work on weed interference and CPWC in cucumber are very limited in the literature, the few available studies [4, 5, 45, 48] have shown that the CPWC in cucumber varies across environment (location, soil, climate and management), infesting plant community (species, density and population), crop (cultivar, spacing and density), growing seasons and years [44–48].

In the USA, the CPWC in cucumber was estimated to be between 4 and 6 weeks after sowing [49–51]. In another study, it was determined that cucumber maintained weed-free for as little as 2 weeks after sowing (WAS) produced yield similar to the season-long weed-free treatment [52]. The author found that a single weeding either 3 or 4 WAS was sufficient to prevent yield loss for cucumber planted on 1.2 m row spacing, and concluded that no CPWC existed. Conversely, cucumber in a narrow row spacing had a 3 to 4 WAS CPWC [52]. It was reported that the CPWC for cucumber was longer at a higher plant population than in a lower plant population [52]. In Canada, the CPWC for cucumber was determined to be between 12 to 36 days after sowing (DAS) with a mixed population of common ragweed and common lambsquarters [45]. In Brazil, it was found that the CPWC was between 3 to 7 WAS [53]. Due to the disparities in the results of the CPWC from one study and location to the other, it has been recommended that critical period of weed interference should be determined specifically for a particular region considering the weed composition and climatic condition in order to provide precise information for growers [54].

5. Weed management in cucumber

5.1 Preventive weed management

Preventive measures of weed control is an important part of weed management that has gained attention among cucumber growers and weed scientist in recent time. Preventive weed control involves techniques and practices that hinders the build-up of weed species [55]. These involves clean cultivation through the use of clean water, seeds and fertilizer, and keeping the farm environment free from weeds and their seeds [56]. It is necessary to begin preventive weed control during the year before the beginning of cucumber production and use cucumber seeds free from weed seeds to promote a weed-free cucumber crop in the preceding season. The selected field must be relatively free from weed species such as nut sedges, Bermuda grass, morning glories and Johnson grass. Seed set by pigweeds, common cocklebur and other aggressive annual weeds must also be avoided as a precautionary measure to achieve a weed-free cucumber field [49]. Other preventive weed management measures in cucumber includes not growing cucumber the year after another annual vegetable with similar tillage cultivation and harvest schedules, especially in a highly infested field. Cultivation fallow can also be used to reduce the weed seed bank in cucumber fields that are heavily infested with weeds. The choice of cucumber variety can also affect the level of weed infestation. Hence, vigorous varieties with good adaptation to the prevailing local conditions and good foliage to suppress weed should be selected [49, 56]. Optimum conditions that give cucumber a competitive advantage over the weed species must be provided to prevent build-up of weed species. Fertilizer application method and timing must be manipulated in such a way that the nutrients are available to the crop rather than the weeds. In-row drip irrigation and fertigation can be used to water and apply nutrients to the cucumber and not the inter-row weeds [42]. Measures should also be taken to use water free from weed seeds. Large amount of rapidly available nitrogen, phosphorus and potassium fertilizers that can stimulate excessive weed growth in nutrient responsive weeds such as pigweeds common cocklebur, common ragweed and lambs quarters should be avoided [42]. Many weed species have higher water use efficiency than cucumber, hence flood irrigation should be avoided because they provide conducive environment for weed to flourish. Furthermore, weeds should be removed before they set seeds to avoid weed seed spread in the cucumber crop. Removing weeds in their early growth stages prevents them from setting seeds and spreading these to other areas of the field. Therefore, it is necessary to remove weeds the first time they have been noticed [42, 49, 56].

5.2 Cultural weed management

Cultural weed control is among the most important means of weed management used easily by most cucumber farmers. Cultural control is the use of common practices such as crop rotation, variation of crop row spacing, competitive cultivars adapted to climate and regional conditions, live mulches, cover crops etc. for the proper management of weeds, water and soil [57]. There has been a growing interest in cultural weed control methods during the last two decades as a result of the increasing concern of pesticide use. Cultural practices are regarded as the second most environmentally friendly weed control method next to preventive measures. Cultural techniques help the cucumber farmers to reduce the cost of weed management. These techniques can affect weed-crop interaction and inter-relationship particularly during the critical period of weed control. These techniques provides favorable and conducive environment for the growth of cucumber and give the crop a competitive advantage over infesting weed species. Cultural weed control methods are easy and cost–effective in cucumber production. Crop rotation, primary tillage, soil solarization, high plant population and manipulation of sowing dates and row spacing are cultural techniques that can easily control weeds in cucumber production [21, 46, 50].

5.2.1 Crop rotation

Crop rotation enhances cucumber productivity by improving weed control and soil productivity. Continuous cultivation and tillage systems have negative interaction with each other and results in a shift in weed species composition with consequence difficulty in weed management [58]. A shift from cucumber to other crops of different life cycle, physiology and morphology serves as an important means of preventive weed control when cucumber is grown over time in the same field [42]. This practice has potential to reduce weed density and biomass, particularly when a competitive crop is rotated and an effective direct weed control tillage system is applied [5, 41]. On the other hand, continues cropping increases the risk of resistant weeds as a result of the application of similar cultural practices and herbicides of same chemistry for longer periods [21]. Broad-leaf weeds which are difficult to control in cucumber and other vegetables can be controlled readily in cereal crops. Crop rotation is particularly important in cucumber production because of its disease control benefit and weed control flexibility [59]. Cucumber-tomato, cucumber-pepper and cucumber-eggplants, rotation in farmers' fields showed that the rotation of cucumber with other vegetable crops is agronomically practicable, sustainable, and an eco-friendly technique for better weed control and economic benefits [59, 60].

5.2.2 Primary tillage

Primary tillage is an integral part of cucumber production system that enhances field preparation for planting operation. The tillage system used directly affect soil structure, plant available moisture and intensity of weed problem. Soil inversion during tillage is considered to be very beneficial for weed control [61]. The implement used and the depth of the tillage operation determines the impact of primary tillage in cucumber farming. The use of moldboard plough is an effective way to reduce weed density during the early growth stages of the cucumber crop [61, 62]. Weed densities and biomass are usually higher in zero or minimum tillage systems than in conventional tillage systems that involves the use mold plough [10, 61–63]. It was reported that cucumbers planted into no tillage rye had greater weed size compared to conventional tillage [61]. Reduced tillage was also reported to encourage increased perennial weed species in weed population in cucumber fields compared with conventional tillage [61, 62].

5.2.3 Stale seedbed

The use of stale seedbed is another cultural practice for suppressing weeds in cucumber. A stale seedbed is defined as a seedbed prepared several days, weeks, or month prior to planting or transplanting a crop [64]. In this method, resurgent weeds in ploughed field are controlled by the use of tillage while irrigation or rain are used to stimulate weed seed germination. The flush of young seedlings is then killed by using shallow tillage or herbicides [65]. This method has been successfully used to reduce competition of several weed species including *Palmer amaranth*, and yellow nutsedge in cucumber [32]. Stale seed bed reduced weed infestation with the applications of glyphosate and paraquat on the seedbeds to control emerged weeds [66].

5.2.4 Soil solarization

Soil solarization is another non-chemical weed control technique in cucumber production. This technique involves hydrothermal disinfection of moist soil by transparent polyethylene sheets during the hot summers. These sheets entrap the sunlight and increase the temperature of upper layers of the soil by 8–12 °C compared with the non-mulch soil. The elevated temperature kills some of the seeds and breaks the dormancy of others. While the solar scorching kills the newly emerged weed seedlings [67]. Soil solarization is a simple, non-hazardous method that avoids the use of any toxic materials, does not contaminate the site and therefore suited for organic cucumber farming. The effectiveness of this method of weed control has been reported in cucumber crop [68, 69]. Soil solarization proved to be an excellent method for complete control of parasitic weed specie such as Egyptian broomrape (*Orobanche aegyptiaca*) and other weed species such as *Sorghum virgatum, Chenopodium album, and Purtulaca oleracea* infestation in cucumber [68–70].

5.2.5 Plant density and row spacing

Manipulation of crop row spacing and planting density can restrict weed seed germination and enhance the crop competitive ability against weeds [71]. Narrow row spacing and high plant densities are important techniques in enhancing cucumber competitiveness and suppressing weed growth [16, 72, 73]. These techniques are very cost-effective and environmental friendly. When the optimum plant population density is used through appropriate row spacing, cucumber crop is able to develop canopy cover and hence competitive advantage over emerging weed seedlings [73]. Narrow row spacing is known to suppress weed growth by closing crop canopy earlier than wide spacing. Early canopy cover by closely spaced cucumber has been shown to smother weeds, hence reducing weed-crop competition [16, 74]. Cucumber planted at narrow plant spacing of 1 m × 0.3 m resulted in earlier canopy closure and better weed suppression than those planted at 1 m × 0.6 m and 1 m × 0.9 m [73]. In another study, spacing of 75 cm × 25 cm resulted in weed density and biomass suppression compared to spacing of 75 cm × 50 cm and 75 cm × 75 cm in cucumber [73]. Herbicides work well with narrow spacing as it impacts the weeds by decreasing their vigor due to high competition with the cucumber plants in narrow row planting compared to the wide row planting [73, 74].

6. Mechanical weed management

Mechanical weed management involves the physical removal of weeds from the field by hand pulling or through the use of farm tools and implements such as hand hoes, cutlasses, cultivators, choppers, mowers disks or weeders [75, 76]. Mechanical weed management is one of the oldest weed control practice. It involves the practices of primary and secondary tillage. With mechanical weed management, weeds in fallow fields are killed and the weed seeds buried in deep soil layers where they cannot emerge. Mechanical weed control in cucumber also involves plowing or disking to destroy weeds by exposing them to variations in light, moisture and temperature [77]. Secondary tillage practices such as harrowing is also used to dislodge and shred weeds in cucumber field. Although these practices destroys weed quickly, they do not provide season-long effect because some weed seeds are still present close to the soil surface [77]. It is therefore imperative to use mechanical weed management before or during early flowering to prevent the production of large quantity of weed seed, and engage follow-up weed control practices to achieve effective weed control. The best practice is usually to cultivate cucumber at the preliminary stage of weed growth when the weeds are still physiologically immature to exert significant competition with the crop [65]. Mechanical weed control cannot be used as the singular method of weed management because it may provide favorable conditions for emergence and dispersal of dormant weed seeds. It also impact the soil structure negatively resulting in soil dryness and compaction [65]. Hence, mechanical weed control must be used only as a supplement to other weed control practices within the context of integrated weed management.

7. Chemical weed management

Chemical weed management in cucumber is mainly through the use of herbicides of different active ingredients. Although only limited herbicides are registered and available for weed control in cucumber, herbicides are an essential component of a successful weed control program in cucumber production [78]. These herbicides either pre-emergence or post-emergence, when applied at correct dosage and appropriate timing hampers weed growth and development [6, 7]. Herbicides use in cucumber reduces drudgery and labor requirement, and makes weed control easy, efficient and economical. It also improves soil structure by boosting soil moisture and reducing soil erosion. However, effective weed control with the use of herbicides is limited by the potential for crop injury from registered herbicides [7, 28,]. Herbicides application at too high rate can damage cucumber while too low rate will not provide the expected weed control [5]. Best results from herbicides application in cucumber are obtained when the weeds are at their highest susceptible stages and the crop is at its highest tolerance stages. Selection of a suitable herbicide program for cucumber depends on the population, growth stage, biology and ecology of the infesting weed species [10, 14, 15]. Much of chemical weed control in cucumber revolves around two key herbicides: ethalfluralin and clomazone which gives a reasonable control over most weeds [20]. Both herbicides are safe to use on cucumber and are generally applied pre-emergence for grass and broadleaved weed control [34]. Pre-emergence and post-emergence herbicides are used for effective weed control in cucumber. Pre-emergence herbicides can be applied before the planting of cucumber. These herbicides remain active in soil and provide control of weeds before they emerge. However, pre-emergence herbicides should be used with extreme care as they can damage the cucumber seedlings [5, 20]. Although pre-emergence herbicides such as N-1-naphthylphthalamic acid (naptalam) provides satisfactory control of the grasses and broadleaved weeds, erratic performance of the herbicide was observed in cucumber [79, 80]. Cucumbers were tolerant to 3.4 to 4.5 kg/ha of naptalam applied immediately after seeding but were injured by applications at emergence or vining [79, 80]. Reduced yields and crop injury with pre-emergence applications of CDEC at 4.5 kg/ha was also reported in cucumber [79, 80].

Registered herbicides for broadleaf weed control in cucumber include halosulfuron, clomazone, ethalfluralin, bensulide, paraquat, carfentrazone and glyphosate. Glyphosate, paraquat, and carfentrazone are effective on *Palmer amaranth* when applied as post-emergence herbicide. However, these herbicides are only registered in cucumber for non-selective control of emerged weeds pre-plant, pre, or post along the crop rows with the use of spray guard [81–83]. These herbicides lack residual control and have limitations when applied post-directed [20], including the failure to control weeds beneath or closes to the crop canopy. Therefore, additional post herbicides that are non-toxic to the crop would be beneficial. Halosulfuron is registered for pre and post-emergence control of some *Amaranth* species [84] but does not give an effective post-emergence control of *Palmer amaranth* [85]. Clomazone has poor efficacy on *Palmer amaranth* when applied alone [85, 86]. Ethalfluralin applied as pre-emergence herbicide provides good early season control of *Palmer amaranth* [87]. Bensulide is an herbicide used as pre-emergence in cucumber and can be tank-mixed with naptalam. Bensulide primarily controls annual grasses, with suppression of only three broadleaf weeds [88]. Bensulide may persist in the soil for months, which may result in potential injury to cucumber [34].

Farmers also often use a combination of clomazone and ethalfluralin for weed management in their cucumber production. Clomazone applied alone suppresses several annual broadleaf weeds and grasses. Clomazone controls*galinsoga* species (Galinsoga spp.), common lambsquarters (Chenopodium album L.), spurred anoda (Anodacristata L.), and velvetleaf (Abutilon theophrasti Medicus.) [20]. However, the herbicide has potential to injure cucumber and adjacent vegetation as a result of volatilization and drift. Ref. [46] found that clomazone caused chlorosis in cucumber plants, though recovery was rapid. Similar to clomazone, ethalfluralin provides efficient control of many broadleaf and grass weeds and may injure cucumber. Carpetweed (*Mollugo verticillata* L.), common lambsquarters, pigweed spp. (*Amaranthus* spp.), common purslane (*Portulaca oleracea* L.), and annual grasses are controlled by ethalfluralin. Injury to cucumber from ethalfluralin differs from that of clomazone in that stunting of plants and thinning of plant stand may occur. A major factor that increases injury from ethalfluralin in cucumber is rainfall, irrigation and increased seeding depth [34, 88]. Combination of clomazone and ethalfluralin provided excellent control of annual grass and broadleaf weeds. Ref. [46] reported that applying clomazone and ethalfluralin together controlled hairy nightshade (*Solanum sarrachoides* Sendt.), redroot pigweed (*Amaranthus retroflexus* L.) and smartweed (Polygonoum persicaria L.) better than either herbicide alone. Although this herbicide combination is effective against a number of weed species, they have little to no activity on weed species such as smooth pigweed (*Amaranthus hybridus* L.), morning glory species (*Ipomoea* spp.), and yellow nut sedge (Cyperus esculentus L.) weed species which interferes with harvesting and reduce cucumber fruit quality [26].

8. Integrated weed management in cucumber

Integrated weed management is the major component of a sustainable cucumber farming. Considering the diversity of weed problem, no single method, whether physical, mechanical or chemical can provide the desired level of efficiency under all situation [19]. Hence, cucumber growers should focus on adopting integrated weed management system to widen weed control spectrum and efficiency in a sustainable, economical, and environmental manner. Integrated weed management involves coordinated use of multiple tactics for optimizing the control of all classes of weed in an ecological and economical sound manner [43]. These tactics can be direct weed control through physical (manual and mechanical tillage/land preparation), chemical and biological means [6]. It could also be indirect control through cultural or agronomic practices such as planting pattern, fertilization timing and placement method, sowing time, row spacing, seed rate, crop cultivar type, intercropping and cover crops [7]. These methods can influence either weed density (i.e. the number of individuals per unit area) and/or weed development (biomass production and soil cover). It is always recommended to use all available options in combination to achieve better control of weeds.

9. Conclusions

Cucumber is a difficult crop to manage as it is susceptible to the attack of numerous weeds, disease pathogens, and insect pests. Weeds reduces cucumber yield and deteriorate fruit quality. Unfortunately, cucumber is not a strong competitor against weeds particularly during the early growth stage. Hence, it is necessary to control weeds to obtain increased yield and high quality fruit from cucumber. All the weed control methods have their own shortcomings and cannot be used as a stand-alone tactics to manage weeds in cucumber efficiently. Manual weed have the constraint of high cost and labor shortage, mechanical options have their own limitations because of the increase in fuel cost, and their use is not practicable within cucumber rows and on large farm sizes. Chemical control on the other hand are always expressive, and only a few herbicides are registered for weed control in cucumber. Moreover, the few available herbicides cannot control the entire weed spectrum and provide season-long weed control when used alone. No single weed control method can provide 100% control; therefore, there is a need to adopt an integrated weed management approach to control weeds in cucumber. A good tillage operation and land preparation, the use of a competitive cucumber cultivar and appropriate plant population and row spacing, application of pre-emergence herbicides, application of post-emergence herbicides particularly along crop rows with the use of spray guard are important in reducing weed density. The combination of these approaches provides effective weed control, improves fruit quality, and helps in environmental conservation. The world is now moving toward precision weed management techniques which involve remote sensing, modelling and use of robotics to control weeds. These technologies are the future of weed management in crop production and have a substantial role to play in modern cucumber production systems. Right selection of one or more of these techniques with reference to environmental, socioeconomic, and geographic conditions will provide effective weed control in cucumber.

Author details

Olumide Samuel Daramola
Department of Plant Physiology and Crop Production, Federal University of Agriculture Abeokuta, Ogun State, Nigeria

*Address all correspondence to: olumidedara01@gmail.com

References

[1] Daramola OS, Adigun JA, Adeyemi OR. Efficacy and economics of integrated weed management in Chilli pepper. Journal of Crop Improvement. 2021;35(1): 38-50.

[2] Chauhan BS, Singh RG, Mahajan G. Ecology and management of weeds under conservation agriculture: A review. Crop Protection. 2012;38:57-65.

[3] Daramola OS. Timing of weed management and yield penalty due to delayed weed management in soybean. Planta Daninha. 2020;38(3):1-7

[4] McGowen SJ, Jennings KM, Chaudhari S, Monks DW, Schultheis JR, Reberg-Horton C. Critical Period for Palmer Amaranth (*Amaranthus palmeri*) Control in Pickling Cucumber. Weed Technology. 2018;32:586-591. DOI: 10.1017/wet.2018.58

[5] Berry AD, Stall WM, Rathinasabapathi B, Macdonald GE, Charudattan R. Smooth pigweed (*Amaranthus hybridus* L.) and livid amaranth (*Amaranthus lividus*) interference with cucumber (*Cucumis sativus*). Weed Technology. 2006;20:227-231.

[6] Webster TM. Weed survey–southern states: vegetable, fruit, and nut crops subsection. In: Proceeding of South Weed Science Society 2006;63:246-257.

[7] Holmes GJ, Monks DW, Schultheis JR, Sorensen KA, Thornton AC, Toth SJ. Crop profile for cucumbers in North Carolina. 2005. https://ipm.ces.ncsu.edu/ipm-crop-profiles/.

[8] Daramola OS, Adigun JA, Adeyemi OR, Adejuyigbe CO. Effect of row spacing and weed control methods on weed population dynamics in soybean. International Journal of Pest Management. 2020b;67(1):1-17

[9] Daramola OS, Adigun JA, Adeyemi OR, Olorunmaiye PM. Efficacy and economic comparison of weed management in tomato. International Journal of Vegetable Science 2020a;26(5):18-25

[10] Ugwunna RU, Ochekwu EB, Anyanwu DI. Effectiveness of the Primextra Gold in controlling weeds of Cucumber (*Cucumis Sativus* L.). Journal of Applied Science and Environmental Management.2015;98

[11] Trader BW, Wilson HP, Hines TE. Halosulfuron helps control several broadleaf weeds in cucumber and pumpkin. Weed Technology. 2007;21:966– 971

[12] Tickes B. Arizona Veg IPM: Melon weed control. Retrieved from Farm Press 2012: http://www.westernfarmpress.com

[13] Sosnoskie L, Hanson B, Lanini WT. Melon tolerance and weed control with new herbicides. California: University of California 2013.

[14] Horak MJ, Loughin TM. Growth analysis of four Amaranthus species. Weed Science. 2000;48:347-355

[15] Shiboleth YM, Arazi T, Wang Y, Gal-on A. A new approach for weed control in a cucurbit field employing an attenuated poty virus vector for herbicide resistance. Journal of Biotechnology. 2001;37-46.

[16] Larry DK and Herbert JH. Plant Density and Herbicides Affect Cucumber Productivity Journal of America Society of Horticultural Science. 2000;117(1):48-53. 1992.

[17] Nweke IA, Orji EC, Ijearu SI. The effect of staking and plant spacing on the growth and yield of cucumber (*Cucumis stativus* L.) Journal of

Environmental Science, Toxicology and Food Technology. 2013;3:2319-2402.

[18] Ekwu LG, Utobo EB, Onyesola CA. Vegetative and Yield Response of Cucumber to Staking and Nitrogen Fertilizer Application. Applied Science. 2007;19(4):7509-7519.

[19] Swanton CJ, O'Sullivan J, Robinson DE. The critical weed-free period in carrot. Weed Science. 2010;58:229-233

[20] Anonymous. Crop Protection Reference, 18 ed. C & P Press, New York 2002.

[21] Tursun IE, Akinci MS, Ahmet U. Estimating Time of Weed Emergence in Cucumber (*Cucumis sativus* L.) Nihat Turkish Journal of Agriculture - Food Science and Technology. 2015;3(5): 271-278.

[22] Wiro KO, Iyagba AG. Influence of weeding frequency and poultry manure on the growth and yield of cucumber (*cucumis sativus*) in the humid region of Nigeria International Journal of Advanced Scientific Research. 2020;5(5):9-15

[23] Tomlinson JA, Anne LC. Weed plants as sources of cucumber mosaic virus. Annals of applied biology. 1970;66: 11-16

[24] Webster TM. Weed survey—southern states: vegetable, fruit and nut crops subsection. In: Proceeding of South Weed Science Society 2000;55:237-258.

[25] Poston DH, Wilson HP, Hines TE. Imidazolinone resistance in several Amaranthus hybridus populations. Weed Science. 2000;48: 508-513.

[26] Liebl RA, Norman MA, Mechanism of clomazone selectivity in corn (*Zea mays* L.), soybean (*Glycine max*), smooth pigweed (*Amaranthus hybridus*), and velvetleaf (*Abutilon theophrasti*). Weed Science. 1991;39:329-332.

[27] Ngouajio MG, Wang G, Hausbeck MK. Changes in pickling cucumber yield and economic value in response to planting density. Crop Science. 2006;46:1570-1575

[28] Ando K, Grumet R. Evaluation of altered cucumber plant architecture as a means to reduce Phytophthora capsici disease incidence on cucumber fruit. Journal of American Society of Horticultural Science. 2006;131: 491-498.

[29] Tepe I. Yabancı otlarla mücadele. SİDAS Medya ltd. Şti. VAN 2014.

[30] Aspiazu I, Sediyama T, Ribeiro Jr, Silva AA, Concenco G, Galon, L, Ferreira EA Araujo WF. Characteristic associated to photosynthesis and water of weed species. Planta Daninha. 2010;10:347-353

[31] Daugovish O, Thil DC, Shaft B. Modelling competition between wild oat and yellow mustard Weed science. 2003;34: 102-109

[32] Johnson III WC, Mullinix, Jr. BJ. Cyperus esculentus interference in Cucumis sativus. Weed Science. 1999;47:327-331.

[33] Altland J. Weed control in container crops: A guide to effective weed management through preventive measures. Oregun State University Extension Service, 2003;8.

[34] Grey TL, Bridges DC, NeSmith DS. Tolerance of curcubits to the herbicides Clomazone, Ethalfluralin and Pendimethalin. II. Watermelon. Hort Science. 2000;35: 637-641.

[35] Jerzykiewicz J, Klobus G. Effect of triazine and urea-type herbicides on photosynthetic apparatus in cucumber leaves. Acta Societatis botanicorum poloniae. 2007;76(1) 43-47

[36] Nagle J. Melon thrips. Department of Employment, Economic Development and Innovation {Qld} 2008;http://www.dpi.qld.gov.au/4790_11215htm

[37] Webb E. Avoid the silver lining. Vegetables Australia July/August: 2008;24-25

[38] McDougall S. Primefacts: Silverleaf whitefly in vegetables. Industry and Investment NSW 2009

[39] Watson A, Napier T. Primefacts: diseases of cucurbit crops. Industry and Investment NSW 2009 http://www/dpi.nsw.gov.au/_data/assets/pdf_file/0003/290244/diseases-curcurbit-vegetables.pdf

[40] Adam KL. Squash bug and squash vine borer: Organic controls [Online]. ATTRA Publication IP298. National Sustainable Agriculture Information Service 2006. Available at: https://attra.ncat.org/attra-pub/summaries/summary.php?pub=138

[41] Webster TM. Weed survey – Southern states. Vegetable, fruit and nut crops subsection. Proceedings of the Southern Weed Science Society 2006;59: 260-277.

[42] Webster, T. M. 2006. Weed survey – Southern states. Vegetable, fruit and nut crops subsection. Proceedings of the Southern Weed Science Society 59: 260-277.

[43] Knezevic SZ, Evans SP, Blankenship EE, Van Acker RC, Lindquist JL. Critical period for weed control: The concept and data analysis. Weed Science. 2013;50:773-786

[44] Ngouajio M, Mennan H. Weed populations and pickling cucumber (Cucumis sativus) yield under summer and winter cover crop systems. Crop Protection 2005; 24:521-234

[45] Friesen GH. Weed interference in pickling cucumbers (Cucumis sativus). Weed Science. 1978;26:626-628.

[46] Alkhaatib K, Kadir S, Libbey C. broadleaf weed control with clomazone in pickling cucumber (Cucumis sativus). Weed Technology, 2005;9:166-172.

[47] Horak MJ, Loughin TM. Growth analysis of four Amaranthus species. Weed science. 2000;48(3): 347-355

[48] Meyers SL, Monks DW, Schultheis JR, Jennings KM. Interference of Palmer amaranth (Amaranthus palmeri) in sweet potato. Weed Science. 2010;58:199-203

[49] Peet M. Sustainable practices for vegetable production in the South. Focus Publishing, R. Pullins Company, Newburyport, MA. 1996.

[50] Noble D. Working toward better cucurbits: Weed management strategies for the coming season 2009 [Online]. Available at: http://www.growingmagazine.com/fruits/working-toward-better-cucurbit

[51] Stall WM. Weed control in cucurbit crops (muskmelon, cucumber, squash, and watermelon 2009. Available at http://edis.ifas.ufl.edu/wg029.

[52] Weaver SE. Critical period of weed competition in three vegetable crops in relation to management practices. Weed Research. 1984;24(5)317-325.

[53] William RD, Warren GF. Competition between purple nutsedge and vegetables. Weed Science. 1975;23:317-323.

[54] Schultheis JR, Wehner TC, Walters SC. Optimum planting density and harvest stage for little-leaf and normal-leaf cucumbers for once-over

harvest. Canadian Journal of Plant Science. 1998; 78: 333-340

[55] Riar DS, Norsworthy JK, Steckel LE, Stephenson IV DO, Eubank TW, Bond J, Scott RC. Adoption of best management practices for herbicide-resistant weeds in midsouthern United States cotton, rice, and soybean. Weed Technology. 2003;27:788-797.

[56] Connolly B. Saving cucurbit seed. The Natural Farmer 2005;2(65): 13-15. Available online at http://www.nofa.org/tnf/pdfs/Summer2005.pdf (verified 18 July, 2011).

[57] Maynard DN, Hochmuth GJ. Knott's handbook for vegetable growers, 1997;4th ed. John Wiley and Sons, New York.

[58] University of California at Davis. Cucurbits: Integrated weed management. 2008: http://www.ipm.ucdavis.edu/PMG/r116700111.html.

[59] Xiao X, Cheng Z, Meng H, Khan MA and Li H, Intercropping with garlic alleviated continuous cropping obstacle of cucumber in plastic tunnel. Acta Agricultura Science. 2012;62:696-705.

[60] Ando K, Grumet R. Evaluation of altered cucumber plant architecture as a means to reduce Phytophthora capsici disease incidence on cucumber fruit. Journal of American Society of Horticultural Science. 2006;131:491-498.

[61] Wang G, Ngouajio M. Integration of cover crop, conservation tillage, and low herbicide rate for machine-harvested pickling cucumbers. Horticultural Science. 2008 43:1770-1774

[62] Swaider M, Ware GW, MacCollum S. "Commercial Cucumber Production." B's Cucumber Pages. Interstate Publishers Inc., 5 Sept. 2015;17.

[63] Uko I, Onunwa AO, Ndukwe PC, Ezennia OL. Soil physico-chemical properties, cucumber (*cucumis sativus* l.) growth and yield as influenced by land preparation and weed control methods . Nigerian journal of crop science. 2019;6(1):14-20.

[64] Lonsbary SK, O'Sullivan J, Swanton JC. Stale seedbed as a weed management alternative for machine-harvested cucumber. Weed technology. 2009;17: 724-730

[65] Dogan MN, Jabran K, Unay A. Integrated weed management in cotton. In: Recent Advances in Weed Management (ed. B. Chauhan and G. Mahajan), 2015;197-222. New York, NY: Springer https://doi.org/10.1007/978-1-4939-1019-9_9.

[66] Noll CJ. Chemical weeding of cucumber in a stale seedbed. Proc. North East Weed Science Society. 1978;32:230-232.

[67] Hosmani MM, Meti SS. Non-chemical means of weed management in crop oduction. Integrated weed management for sustainable agriculture. 1993;1:299-305. In: Proceedings of a conference held at the Department of Agronomy, CCS Haryana Agricultural University, Hisar, Haryana, India.

[68] Ashrafi ZY, Hassan M, Alizadeh H, Sedigheh S. Effect of soil solarization on the control of Egyptian broomrape (Orobanche aegyptiaca) and yield improvement of cucumber (Cucumis sativus) grown in greenhouse. Bulgaria Journal of Agricultural Science. 2008; 14: 583- 591

[69] El-Sayed SF, Abdel-Wahab, A, El-Taweel, H. Effect of Bio-fertilization and Sterilization on Cucumber Production under Plastic House Conditions Middle East Journal of Agriculture. 2016;5(2):186-200

[70] Qasem, JR, Kasrawi, MA. Variation of resistance to broomrape (Orobanche

ramosa) in tomatoes. Euphytica, 1995;81: 109-114.

[71] Eslami, S.V. Weed management in conservation agricultural systems. In: Recent Advances in Weed Management. 2005;87-124. New York, NY: Springer https://doi.org/10.1007/978-1-4939-1019-9_5.

[72] Sunday O, Udensi UE, Nwachukwu JC. Response of Cucumber (*Cucumis sativus L.*) and Weeds to Plant Spacing and Weeding Regimes in a Humid Forest Agro-Ecology of Southeastern Nigeria International Journal of Plant and Soil Science. 2018;22(1):1-13,

[73] Sanni KO, Adenubi OO. Influence of intra row spacing on weed suppression in cucumber (*Cucumis sativus*) production in humid rainforest agroecological zone of lagos, Nigeria. Global Journal of Ecology. 2020;5(1):038-043. DOI: https://dx.doi.org/10.17352/gje.000018

[74] Vikram KK, Ameta KD, Suresh KT, Akshay C, Suman G, Satveer Y. Effect of spacing and training on growth and yield of polyhouse grown cucumber (*Cucumis sativus* L.). International Journal of Microbiology and Applied Science. 2017;6(8):299-304.

[75] Lègère A, Samson N, Rioux R. Perennial weeds in conservation tillage systems: More of an issue than in conventional tillage systems? Proceedings of 1993 Brighton Conference - Weeds. 1993;747-752.

[76] Melander B. Mechanical weed control in transplanted sugar beet. Proceedings of the 4th Workshop of the EWRS Working Group on Physical and Cultural Weed Control, Elspeet The Netherlands, 2000;25.

[77] Seaman A. production guide for organic cucumbers and squash. NYS IPM Publication 2009;135. Cornell University and the New York State IPM Program. (Available online at: http://www.nysipm.cornell.edu/organic_guide/cucurbit.pdf).

[78] Mitchem WE, Monks DW. Cucumber (Cucumis sativus) response to bentazon and halosulfuron. WSSA Abstracts. 1997;37:20.

[79] Romanowski RR, Tanaka JS. An evaluation of herbicides for use with cucumbers (Cucumis sativus) and watermelon (*Citrillus vulgaris*) in Hawaii. Ha waii Agr. Exp. Sta. Progr. Rep. 1965;144:30

[80] Lange AD,. May B, Fisher F, schweers V. Weed control in cucurbits. Calif. Agr. 1968;22(3):8-9.

[81] Anonymous. Roundup Powermax® herbicide label. St. Louis, MO: Monsanto Company. 2012a;56

[82] Anonymous. Gramoxone® SL herbicide label. Greensboro, NC: Syngenta Crop Protection, LLC. 2011b;55

[83] Anonymous. Aim® EC herbicide label. Philadelphia, PA: FMC Corporation. 2011a;16

[84] Anonymous. Sandea® herbicide label. Yuma, AZ: Gowan Company. 2012b;22

[85] Norsworthy JK, Meister CW. Tolerance of cantaloupe to postemergence applications of rimsulfuron and halosulfuron. Weed Technology. 2007;21:30-36

[86] Meyers SL, Jennings KM, Monks DW. Herbicide-based weed management programs for Palmer amaranth (Amaranthus palmeri) in sweet potato. Weed Technology. 2013;27:331-340.

[87] Shrefler JW, Brandenberger LP, Webber CL III, Roberts W, Payton ME,

Wells LK. POST weed control using halosulfuron in direct-seeded watermelon. Weed Technology 2007;19:706-712

[88] Derr JF Monaco TJ. Ethalfluralin activity in cucmber. Weed science. 1982; 30(5):498-502.

Section 5

Pest and Disease Prevention

Chapter 10

Integrated Pest Management for Cucurbits in Cucumber (*Cucumis sativus* L.)

Ravi Mohan Srivastava and Sneha Joshih

Abstract

The vegetables belonging to family cucurbitaceae are known as cucurbits. These vegetables are attacked by various insect pests right from seeding to harvest. A lot of money, time, and natural resources are invested to cultivate these vegetables. Sustainable pest management practices can save this investment by avoiding losses. Successful cultivation of cucurbits especially cucumber requires an effective and economical control of insect pests. Commercial vegetable growers must produce quality vegetables that are attractive and safe to the consumer at a minimum cost. Insect pest infestations in cucurbits cause heavy economic losses to farmers through reduction in yield, increased cost of production and lowered quality of produce. Effective and economic and sustainable pest management requires the use of cultural, mechanical, biological, and chemical methods. The integration of these different methods is necessary for achieving good management of pests. In case of cucurbits especially for cucumber pest management can be achieved only by a long-term assurance to integrated pest management practices (IPM). IPM involves the strategic use of resistant varieties, monitoring of pest incidence, cultural methods, mechanical removal of pest, biological control, and need based use of selective pesticides. Integrated pest management (IPM) is the alternative to insecticide and facilitates sustainable environment management.

Keywords: Cucurbits, Pest Management, Insect Pests, Whiteflies, Aphids

1. Introduction

Crops belonging to the family Cucurbitaceae are commonly called cucurbits, which include more than 120 genera and 1000 species distributed in tropical and subtropical areas. Cucurbits are tropical in origin and grown mostly in Africa, tropical America, and Asia, mainly Southeast Asia. Cucurbits are a large group that include cucumber, bitter gourd, squash, bottle gourd, ridge gourd, snake gourd, watermelon, muskmelon and pumpkin. Most of the gourds and squashes are commonly used for cooking while cucumber is used for salads and pickles. Watermelon and muskmelon are taken as fruits and wax gourd is prepared as biscuits and jam. The cucurbits are beneficial for human health as their fruits help in purification of blood, boost energy level, improve digestion in the body and remove constipation. *Benincasa hispida* a member of cucurbitaceae family contains volatile oils, uronic acid, carotenes, flavonoids, ß-sitosterin and glycosides which are of pharmaceutical importance [1]. The major phytochemical present in cucurbits is a terpenoid substance known as Cucurbitacins.

Most cucurbit species is believed to have an African origin. However, cucumber have originated from the foothills of the Himalayas. There the closely related wild species *C. hardwickii Royle* still exists. In India, the cucumber was already being cultivated 3000 years ago, in the 6th century, cucumber was cultivated in China and now it is cultivated worldwide. Immature fruits of cucumber are used as salad vegetable and for pickles. Cucumber and other cucurbit fruits are generally fat-free and low in sodium content. Cucumber is an annual crop and a climbing herb. Its growth development duration can last 12–13 months in favourable conditions.

In present scenario, a significant constraint in the sustainable production and productivity of cucurbits is mainly due to attack of various insect-pests which are responsible for adversely affecting the qualitative and qualitative yield. A wide range of pest complex has been noticed infesting the cucurbits. This chapter joins all the information and provide overall review about the major insect pest of cucumber and other cucurbits, and different IPM measures like monitoring, cultural methods, host resistance, biological control method, use of botanicals as biopesticide and at last use of less hazardous chemical control methods recommended for management of this pest.

1.1 Melon fruit fly, *Bactrocera cucurbitae* (coquillett)

Bactrocera cucurbitae is a yellowish brown coloured dipteran fly and commonly known as melon fruit fly. It is native to India and is wildly distributed in tropical, subtropical and temperate region of the world [2]. Melon fruit fly attacks about 125 species of plants, mainly cucurbits, viz. gourds, cucumber, pumpkin, squash and other vegetables like tomato, green beans, egg plants etc. The female adult with the help of its sharp and hard ovipositor, punctures the soft and tender fruits and lays eggs just below the fruits's epidermis. After hatching, the maggots start feeding inside the pulp of fruits. A water soaked appearance of fruits develops as a result of larval feeding, these punctures and feeding tunnels provide entry points for various bacteria and fungi and result of this, the infested fruit start rotting, distorted and malformed fruits from plants pre-maturely. About 1000 white colour eggs in the batches of 1–40 slender (2 mm long) are laid by a single female. Dirty white coloured apodous maggots are cylindrical, elongate (7-12 cm). The maggot developmental period is 3 to 21 days depending on the temperature and host. Pupation occurs in the soil at the depth of 0.5 to 15 cm, pupal period is about 7–13 days depending on host and temperature. Adults are long lived about 150 days [3]. Melon fruit fly can cause devastating fruit damage up to 100% in all cucurbits [4].

1.1.1 Management

Cucurbits fruits are picked up at small intervals for marketing and self-consumption for that reason we cannot rely on insecticide for control of this pest but under severe infestation, it is important to use low residual toxic insecticides with short waiting periods. Therefore in keeping view the importance of crop and this pest different strategy should be used for the management of fruit fly as follows.

1.1.1.1 Local area management

Local area management is the minimum scale of pest management over a restricted area l, which has no natural protection against reinvasion. Here the major objective is to suppress the pest, rather than to eradicate it. The strategies includes bagging of fruits, field sanitation, protein baits and cue-lure traps, host plant resistance, biological control, and soft insecticides, can be employed to avoid environmental and health hazard [4].

1.1.1.2 Monitoring and management with cue/pheromone lures

For the monitoring of fruit flies, the sex attractant cue-lure traps are more effective than the food attractant tephritlure traps. and cue-lure and methyl eugenol traps have been used to attract males for monitoring and mass trapping [5, 6]. *Ocimum sanctum* as the border crop sprayed with protein bait containing spinosad as a toxicant found effective in management of this pest [7]. A variety of commercially produced attractants (cuelure ®, Eugelure®, Flycide®) are available in the market and can be used efficiently in management of this pest. For trapping male flies, installation of old used water bottle baited with cue-lure saturated wood blocks (ethanol/ cue-lure/carbaryl in a ratio 8:1:2) at 25 traps/ha prior to flower initiation is quite effective. Use of NSKE 4% as a repellent can enhance trapping and luring in bait spots. Use of neem as a repellent enhanced the catch in parapheromone traps and increased the luring ability of para-pheromone by 52%. Although, along with repellents and bait spray, other operations like removal and destruction of maggots in early infested fruits and field sanitation must be adopted.

1.1.1.3 Cultural methods

Field sanitation is most effective method in melon fruit fly management. It should be done for minimising pest intensity and to break reproduction cycle by removal and destruction of infested fruits daily from the field and bury the damaged fruits 0.46 m deep into the soil [8]. Akhtaruzzaman et al. [9] recommended bagging of cucumber fruits after 3 days of anthesis and bag should be kept for 5 days for effective control. Bagging of 3–4 cm long fruits with two layers of paper bags reduces fruit fly infestation and enhances the net returns by 40–58% [10]. 2–3 rows of maize as a trap crop can be grown between the cucurbits, which can be used as a resting site by the adult fruit fly. Spraying of any contact insecticides can be done on maize during the evening hours to kill adult fruit flies.

1.1.1.4 Biological control

Srinivasan [11] reported *Opius fletcheri* Silv. to be a dominant parasitoid of *B. cucurbitae.* The parasitization of *B. cucurbitae* by *O. flatcheri* has been reported to vary from 0.2 to 1.9% in *M. charantia* [12]. In the IPM program of *B. cucurbitae* at Hawaii, a new parasitoid, *Fopius arisanus* has been included [13]. A nematode, *Steinernema carpocapsae* Weiser *(Neoaplectana carpocapsae),* has been found to cause 0–86% mortality to melon fruit fly after 6 days exposure [14]. The culture filtrate of the fungus, *Rhizoctonia solani* Kuhn, act as an effective bio-agent against maggots [15], and the fungus, *Gliocladium virens* Origen, has been found to be an effective against adult flies [16]. Oviposition and development of *B. cucurbitae* adversely affected by culture filtrates of the fungi *R. solani, Trichoderma viridae* Pers., and *G. virens* [17].

1.1.1.5 Host plant resistance

Host plant resistance is an important element in IPM programs It is an important component in IPM programs. It does not cause any negative impact to the environment. The success in developing high yielding and melon fruit fly-resistant varieties is limited. By using wide hybridization technique the resistance genes from the wild relatives can be transferred in the cultivated genotypes of cucurbits [4]. Some resistant genotypes are IHR 89, Hisar II for bitter gourd, Arka Suryamukhi, IHR 35 for pumpkin, NB29 and Pusa smooth purple long for bottle gourd, Arka Tinda for round melon, *Cucumis callosus* for Wild melon, NS14 for sponge gourd and NR2, NR5, NR7 for ridge gourd.

1.1.1.6 Botanicals

The adult longevity reduced from 119.2 to 26.6 days by continuous feeding with extract of *Acorus calamus* (0.15%) mixed with sugar (at 1 mL/g sugar) [18]. Neem oil (1.2%) and neem cake (4%) are found to be effecting in fruit fly management [19].

1.1.1.7 Chemical control

The use of chemical insecticide to manage fruit fly is relatively ineffective. Application of malathion (0.05%) as cover spray to kill the insects on contact is effective or a bait spray by adding 50 g gur + 10 mL malathion in 10 L water that attract and kill the adults is good for its management. For good control of fruit fly spraying of malathion + molasses + water in the ratio of 1:0.1:100 is suggested by Akhtaruzzaman et al. [20]. This method is economical and environmental friendly as there is very low contamination of fruits from insecticides. Gupta and Verma [21] recommended that fenitrothion (0.025%) in combination with protein hydrolysate (0.25%) minimised melon fruit fly infestation to 8.7% compared to 43.3% damage in untreated control. Reddy [22] reported that triazophos is the most effective insecticide to manage this pest on bitter gourd. Diflubenzuron is also found effective in controlling the melon fruit fly [23], and quinalphos (0.2%) successfully manage its population on different cucurbits.

1.1.2 Wide area management

The objective of wide area management is to combine and coordinate different characteristics of an insect eradication program over an entire area within a defensible perimeter. This management program includes a three-tier model, that is, early reduction in population using bait spray, inhibition of reproduction using para-pheromone lure blocks to eradicate males to avert oviposition by females, and intensive survey using traps and fruit inspection until it could be discovered that the pest is entirely eradicated [24]. Male-sterile technique, also incorporated under this program, in this sterile males are released in the fields for mating with the wild females. The transmission of dominant lethal mutations kills the progeny of fruit fly. The females either do not lay eggs or lay sterile eggs. Eventually, the pest population can be eradicated by maintaining a barrier of sterile flies [4]. The spread of the melon fruit fly can be blocked through tight quarantine and treatment of fruits at the import/export ports.

1.2 Red pumpkin beetle (*Raphidopalpa foveico* llis/*Aulacophora foveico* llis)

Red pumpkin beetle is a common and major pest of a wide range of cucurbits, especially sweet gourd, bitter gourd, watermelon, bottle gourd, and muskmelon. It is polyphagous in nature [25]. Both larval and adult stages are voracious feeders of the cucurbit crops and cause severe damage to seedlings and young and tender leaves and flowers [26, 27]. The pest is widely distributed in different parts of the world, especially in Asia, Africa, Australia, and south Europe. In India, it occurs throughout the country but is more common in the north western parts.

The dorsal body of the adult beetle is deep orange in colour, while the ventral side is black. The beetles are about t 5–8 mm in length and 3.5–3.75 mm in width. The posterior part of the abdomen bears soft white hairs. The female lays brown coloured elongated eggs in the moist soil in the clusters of 8–9 that hatch into larvae in 6–15 days. After about 7 days of emergence, beetles starts egg laying and complete its 5 generations from March to October. Larvae are creamy, yellow-coloured and feed on the roots, stems, and fruits touching the ground. Due to infection by the saprophytic fungi rotting of damaged roots and underground stems may be occurred. Leaf lamina of cucurbits is voraciously feed by scrapping off the chlorophyll and making irregular holes on leaves with netlike appearance by beetle. First generation is more injurious than the subsequent generations as the maximum damage is noticed during cotyledon stage. The infested plants may shrivel, and resowing/replanting may become important under severe infestation. The young and smaller fruits of the infested plants may dry up. Sweet gourd was the most susceptible while bitter gourd was found to be the least susceptible host for this pest [28]. Beetles are strong fliers and very active in hot weather, and when disturbed, take fly quickly. If it is not controlled timely the damage by this pest may become severe. The losses due to the pest have been reported up to 30–100% in the field conditions [29].

1.2.1 Management

1.2.1.1 Monitoring and cultural methods

Monitoring of seedling should be done twice a week to check the infestation of red pumpkin beetle. The older plant parts should be monitored regularly and should be treated if severe defoliation is noticed. In the initial stage of infestation collection and destruction of beetles is good practice, otherwise the alternative practices can be employed for the management of this pest. Preventive measures such as burning of old plant parts, ploughing, and harrowing of field after harvesting of the crops should be followed for the destruction of all the stages of this pest. Early planting of cucurbits than the normal planting time could also be effective in management as the plants pass the cotyledonary stage by the time the beetles become active.

1.2.1.2 Host plant resistance

Khan [29] reported the preferred cucurbit host plants for this beetles and categorised bitter gourd, ribbed gourd, sponge gourd, and snake gourd as non-preferred hosts (resistant), while cucumber, muskmelon, and sweet gourd were the most preferred hosts (susceptible) and bottle gourd and ash gourd were moderately preferred hosts (moderately susceptible).

1.2.1.3 Botanicals and biopesticides

Application of neem oil cake in the soil is effective in killing grubs. Treatment of bottle gourd plants with entomopathogenic fungi *B. bassiana* resulted in maximum reduction of beetle infestation along with the highest fruit yield [30]. Khan and Wasim [31] reported maximum repellence against pumpkin beetles in treatment comprising of neem extracts mixed with benzene. *Parthenium spp*. extract was found to be effective in management of the red pumpkin beetle [32].

1.2.1.4 Chemical control

Generally, this pest attacks the crop at the cotyledonary stage when adults skeletonize the young leaves. At the time of initial infestation, applications of malathion (0.5%) or carbaryl (0.1%) minimise the damage successfully [33]. Dusting with permethrin (0.5%) alone and ash + permethrin dust (2000:1 *a.i.* w/w) provide good control against this beetles on the cucumber crop, with no phytotoxicity symptoms on the plant [34]. Synthetic pyrethroids (cypermethrin 0.012%, deltamethrin 0.004%, and fenvalerate 0.01%) were effective in suppressing the beetle population for about a week [35].

1.3 Hadda beetle (*Epilachna implicate, E. viginitioctopunctata and E. borealis*)

This is an occasional pest on cucurbit. It is a serious pest of squash, bitter gourd, and pumpkin. Grub only feed on the underside of the leaf, whereas adults may be found to feed on both the leaf surfaces or even on the fruit rind, leaving spiral-shaped scars and deteriorating fruit quality [36]. Both grub and adult feed voraciously by scrapping the chlorophyll of the leaves which cause characteristic skeletonization of leaf lamina and leaving a fine net of veins on leaves. The affected leaves gradually dry and fallen down. The young plants are killed overnight in case of severe infestation. The yellowish-brown coloured adult beetles are globular in shaped and 6–8 mm long, bearing 12–28 black spots on the elytra. The female lays about 300–400 eggs in clusters on the under surface of the leaves. Eggs on hatching turns into yellowish larvae (grub) with branched black spines covering the body. Full-grown larvae about 7-9 cm long pupate below the leaf or at the base of the stems. The pupa is yellow in colour, and lacks spines and hangs from the leaf. The development period comprises of 4–6 weeks under optimal conditions. Adults make semicircular cuts in rows, while Scrapping of the epidermis indicates the feeding manner of the grubs. Young plants can be totally destroyed, but older plants can tolerate considerable leaf damage. Overwintering sites for adults are under loose tree bark or under leaf litter near the edge of fields.

1.3.1 Management

1.3.1.1 Cultural methods

The best time of day for the inspection of hadda beetles on cucurbits is around noon. As they are not strong fliers, so crop rotation to distant fields tends to limit population and colonisation [36]. On the small scale like kitchen garden handpicking is recommended for the management of this pest because both the larvae and adults are not very aggressive defoliators. Harrowing and destroying vines and larvae after harvesting early cucurbits is recommended to suppress pest population. Row covers can be used to protect cucurbit from the beetles.

1.3.1.2 Botanicals

The application of seed extracts of *Annona squamosa* (3 mL/L of water) helps in minimising population build up to the extent of 76% which is followed by 64% and 57% through NeemAzal (5 mL/L of water) and petroleum ether rhizome extracts of *A. calamus* (2 mL/L of water), respectively [37]. Aqueous neem kernel extracts foliar spray at concentrations of 25, 50, and 100 g/L once in a week and neem oil spray with an ultralow-volume sprayer at 10 and 20 L/ha significantly was found effective in reducing feeding by *Epilachna* beetles in cucumber and squash [38]. Tephrosia leaf extract (20 g/100 mL water) provide highest yield and good control of *Epilachna* beetle by killing adults and inhibition of pupae formation and this is an environment friendly pest control method [39]. Swaminathan et al. [40] have observed the antifeedant and lethal effects of *Azadirachta indica*, *Madhuca latifolia* and *P. glabra,* on this pest. Islam et al. [41] performed larvicidal bioassays with crude aqueous leaf extracts of *plants viz., Ricinus communis, Datura metel* and *Calotropis procera* and these extracts showed considerable toxicity against the Hadda beetles by adversely affecting both oviposition and egg hatching besides prolonged larval duration, pupae formation, and adult emergence.

1.3.1.3 Chemical control

Foliar applications of some synthetic pesticides like parathion, malathion, lambda-cyhalothrin, pyrethrin and spinosad are found to be effective in control of severe infestation of this pest.

1.4 Aphid (*Myzus persicae and Aphis gossypii*)

Many aphid species including melon aphid (*Aphis gossypii*) and green peach aphid found to feed on cucurbits and cause similar devastating damage. They pierce the tender plant part with their slender mouthpart and suck plant fluids from it. Aphids are small about 3 mm long, pear-shaped soft bodied insects with long legs and antennae. They are yellow, pale green, red, brown or black in colour. Some aphid secretes a waxy grey or white material that covers their body and it gives them a waxy or woolly appearance. The adults are usually wingless but when populations are high especially during spring and autumn season winged forms can also be seen. A pair of tube-like structures known as cornicles projecting rearwards from their abdomen is present in most of the aphid species. Aphids can disperse long distances with the help of wind flow. Asexual reproduction is a common phenomenon in majority of aphid species. Adult females give birth to wingless nymphs which become adults after moulting and shedding the skin multiple times within a week. Each adult reproduces numerous nymphs in a short span of time therefore, aphid population increase rapidly. The green peach aphid (*Myzus persicae*) is slender, dark green to yellow in colour, and it has no waxy bloom. They tend to cluster on succulent plant parts and within 10–12 days one generation completes and there are about over 20 generations annually under mild climates [42]. This aphid (both nymph and adult) is known as the most important vector for the transmission of viruses throughout the world [43]. Aphids' infestation causes a variety of symptoms, including reduced plant growth and vigour, yellowing, mottling, browning, curling and wilting of leaves, Ultimately result in low economic yields and sometimes death of plant. The downward curling and crinkling of the leaves is the first sign of aphid infestation. Malformed flowers or fruits are developed due to feeding of aphid on flower buds and fruits. Honeydew excreted by aphids also act as a growth substrate for sooty moulds (fungi) on leaves and other plant parts, which ultimately hinders photosynthesis by blocking light. Aphids also transmit several viruses that affect all cucurbits causing a high rate of crop failure and great economic losses. These viruses are cucumber mosaic virus, zucchini yellow mosaic virus, watermelon mosaic virus and papaya ring spot virus. Mottling, yellowing, or curling of leaves and stunted plant growth are some usual symptoms of viral infection.

1.4.1 Management

1.4.1.1 Monitoring

Aphids multiply at a very fast rate, and this must be considered while monitoring this pest. Plants should be inspected at least two times a week especially at the under surface of the leaves. Severe infestation is noticed during end of growing season. Yellow sticky traps should be used for detecting aphids 2–3 weeks prior to planting.

1.4.1.2 Cultural method

Floating row covers or reflective mulches may be effective to exclude or repel this pest [43]. On early crop stage, aluminium foil mulches can be used to repel the invading aphid and check transmission of viruses. Reflective mulches can increase temperature beyond optimum for cucurbits in very hot and arid regions therefore, not recommended for these areas. Reflective mulches repels aphids from plants and consistently suppress aphid population and also help in delaying symptoms of watermelon mosaic and cucumber mosaic cucumo virus and zucchini yellow mosaic potyviruses by 3–6 weeks. Biodegradable synthetic latex spray mulches and reflective polyethylene and provide good control of aphids and aphid-borne virus diseases on late-season melons [44]. Living mulches minimise the contrast between the plant foliage and bare land, subsequently, the aphids do not detect their host and these mulches around main crop provide additional feeding sites for viruliferous aphids (aphids carrying virus) and hence minimise aphid-borne non -persistently transmitted viruses incidence and spread [45].

1.4.1.3 Biological control

Beneficial insects like predators, parasitoids, and pathogens are attracted to plants with moderate to heavy aphid infestations. These natural enemies may attack large numbers of aphids but as the reproductive potential of aphids is very high hence, the impact of the these natural enemies may not be enough to keep the aphids population below economic threshold levels. Lady bird beetles and their larvae *(Hippodamia convergens, Harmonia axyridis, Coleomegilla maculata)* [46], larvae of the syrphid fly [47], minute pirate bug *(Orius tristicolor* and *O. insidiosus)* and green lacewing larvae (*Chrysoperla carnea, Chrysopa spp.*) and brown lacewing larvae *(Hemerobius spp.)* and larvae of the aphid midge *(Aphidoletes aphidimyza*) [48] are common predators of aphid and help in natural control of this pest. Parasitoids comprises important place among natural enemies of aphid [49]. Some commonly found aphid parasitoids are *Aphidius matricariae, Aphidius colemani, Binodoxys angelicae and Lysiphlebus fabarum* [50]. Under humid conditions, Some fungi infect and provide biological control of aphid population, the most common entomopathogenic fungi are *B. bassiana, M. anisopliae, Verticillium lecanii* [51], and *B. bassiana* must be applied three times at an interval of 5–7 days for good control. Entomopathogenic fungus could be more effective than insecticides for controlling large populations of aphids if utilised properly.

1.4.1.4 Chemical control

Potassium soap and petroleum oil or Actara are recommended for the management of this pest. Killing of aphid population should be done before destroying old crops to avoid winged virus-infected aphids from getting to nearby crops by using a detergent and vegetable oil solution. Acetamiprid (0.01%) or Cypermethrin (0.01%) or malathion (0.05%) can be used to control aphids.

1.5 Whitefly (*Bemisia tabaci, B. argentifolii and Trialeurodes vapararioru*s)

Cucurbits are attacked by several species of whiteflies. Of these, silverleaf whitefly *(Bemisia argentifolii)*, tobacco whitefly *(Bemisia tabaci)*, and greenhouse whitefly *(Trialeurodes vaporariorum)* are the most devastating. Whiteflies are small in size about 1–1.5 mm long and the body and wings of adult are covered with fine whitish powdery wax. Adults and eggs are commonly found on younger leaves while nymphs are present on older leaves. A female lays around 300 eggs [52]; eggs are oval in shape and are laid by making a slit in the leaf. Initially the eggs are white, changing to brown colour, and are hatched within 8–10 days. The first instar is known as called crawler is the only mobile instar that moves to look for feeding sites, while the other instars are sessile and complete its life cycle on the same feeding site [53]. One generation of whitefly completes in about 3–4 weeks. The silverleaf whitefly gets injects a toxin into the plant that causes whitening of the under surface of newly emerging leaves. Severe damage may be occur on younger plants compared to older plants. Whitefly directly affects the cucurbits by its feeding and by acting as a vector of viruses. Cucurbit yellow stunting disorder virus, Cucurbit chlorotic yellow virus, beet pseudo yellow virus, and lettuce infectious yellows virus, are exclusively transmitted by whiteflies in both field and greenhouse-grown cucurbits [54]. Whiteflies also excrete honeydew which promotes the growth of sooty mould on leaves and economic plant parts.

1.5.1 Management

1.5.1.1 Monitoring

For early detection and monitoring the activity of whiteflies in the field, yellow sticky traps can be used. These traps are very important step for their management of whitefly. Traps should be placed just above the canopy at every 100 m^2 distance of the crop as whiteflies are most attracted toward young foliage [55].

1.5.1.2 Cultural methods

Crop rotation, mulching, floating row covers, non-infested transplants, cover crops, and good field sanitation are some common cultural practises to prevent the build-up of whiteflies. Delayed planting or host-free periods may reduces severity of infestation as temperature and rainfall influence whitefly population dynamics [56, 57]. Soil ground covers such as living or synthetic mulches have been found to suppress whitefly infestation [52, 58]. UV-reflective plastic mulch provide good management of silverleaf whitefly by repelling the adults flies, which minimize their colonisation and nymph population on zucchini squash and pumpkin [59]. Field sanitations an important practice to control whiteflies infestation, whitefly-transmitted virus incidence, and insecticide resistance. Crop residues that shelter whiteflies should be destroyed immediately after final harvesting to reduce their population and sources of plant viruses [60]. Weeds should be eradicated regularly from the crop as they can support large populations of whiteflies.

1.5.1.3 Biological control

Natural enemies of whitefly such as *Encarsia formosa*, *E. luteola*, and *Eretmocerus californicus* have been found quite effective in the greenhouse conditions [61]. Entomopathogenic fungi based products namely, *V. lecanii*, *B. bassiana* and *Paecilomyces fumosoroseus* are very useful to suppress whiteflies in both field and greenhouse crops [62]. The important predators affecting whiteflies are true bugs (Miridae, Anthocoridae), beetles (Coccinellidae), lacewings (Chrysopidae, Coniopterygidae), spiders (Araneae) and mites (Phytoseiidae), [63]. *Conwentzia africana* (dusky lacewing) has been considered one of the most important predators of *B. tabaci* [64].

1.5.1.4 Botanicals

Azadirachtin containing neem-based pesticide formulations like NeemAzal and Azatin have been found to control young nymphs and inhibit growth and development of older nymphs, and suppress egg laying by adult flies. Soap and certain oil sprays can be used in an organically certified crop. The efficacy of neem-based pesticides can be increased by adding 0.1%–0.5% soft soap in it.

1.5.1.5 Chemical control

The adults at immature stage reside on the underside of leaves, therefore whiteflies are usually difficult to control by using chemicals. Neonicotinoid insecticide like imidacloprid or thiamethoxam applied in soil at the time of planting effectively controls whiteflies. Foliar spray with neonicotinoid insecticide such as acetamiprid can be done at early stages of growth before flower initiation. Spiromesifen is effective against immature stages of the whitefly [60]. Consecutive applications of the same insecticide should be avoided. Soil application of any neonicotinoid with a foliar application of another neonicotinoid never be followed.

1.6 Squash bug (*Anasa tristis*)

The pest has been reported to attack nearly all cucurbits but most preferred cucurbits for oviposition and high rates of reproduction and survival are squash and pumpkin. It damages the crop severely by secreting highly toxic saliva into the cucurbit. Foliage is the primary site of that wilts, becomes blackened, and dies upon feeding. The fruits are also infested. The intensity of damage is directly proportional to the population density of pest. The adults emit a strong odour when crushed.

1.6.1 Management

1.6.1.1 Cultural methods

Plastic and spun bounded material can be used a row cover at the time of planting. Straw mulch is found effective in controlling squash bugs by providing cover [65]. As this pest prefers squash over other cucurbits, squash planting can be used as a trap crop around other cucurbits. Trap crop squash can be treated with insecticide to control the infestation.

1.6.1.2 Biological control

Several natural enemies are known to parasitize squash bug, especially Hymenoptera parasitoids belonging to family Encyrtidae and Scelionidae. A important tachinid fly parasitoid *T. pennipes*, attacks nymphs and adults of this pest [66], and 100% parasitisation in certain fields in particular years was observed [67].

1.6.1.3 Botanicals

Certain plant derived oils such as neem oil are helpful in the management of squash bug.

1.6.1.4 Chemical control

For the successful squash bug control timing of application is the key. Systemic insecticides are effective in suppression of these bugs up to 3 weeks. Foliar sprays targeting newly hatched nymphs are more effective than sprays used against older stages. Multiple foliar sprays are often needed for long periods of control. Soil application of dinotefuran and pyrethrin in foliar application are recommended for management of this pest [68].

2. Conclusion

The attacks made by the insect pests in cucurbits cause severe yield and quality losses in cucurbits. Cucurbitaceous crop are an important part of the fresh market vegetable crops. The current pest management still relies mainly on chemical pesticides and excessive dependence on chemicals leads to environmental pollution, pest resurgence, pest resistance and disturbance in balance between pest and their natural enemies. There are also the real and important risks to human health and environment as insecticide residue persist in these vegetables for longer times. Therefore, an integrated approach including monitoring of pests; cultural methods, like field rotation, use of mulches and trap crops and shifting planting dates; resistant cultivars; biological control; botanicals and biopesticides; and judicious use of chemicals can minimise these associated risk with chemical pesticides. An effective integrated programme for pest management is necessary for the management of these pest problems in cucurbits. By giving focused attention through adopting IPM techniques, sustainable production of cucurbits can be achieved.

Author details

Ravi Mohan Srivastava and Sneha Joshih
Department of Entomology, College of Agriculture, G.B. Pant University of Agriculture and Technology, Pantnagar, Uttarakhand, India

*Address all correspondence to: ravimohanento@gmail.com

References

[1] Al-Snafi AE. 2013. The Pharmacological Importance of Benincasa hispida- A Review. International Journal of Pharma Sciences and Research, 4: 165-170.

[2] Fletcher BS. 1987. The biology of Dacine fruit flies. Annual Review of Entomology 32: 115-144.

[3] Koul VK, Bhagat KC. 1994. Biology of melon fruit fly, Bactrocera (Dacus) cucurbitae Coquillett (Diptera: Tephritidae) on bottle gourd. Pest Management and Economic Zoology 2: 123-125.

[4] Dhillon, M.K., R. Singh, J.S. Naresh, and H.C. Sharma. 2005. The melon fruit fly, Bactrocera cucurbitae: A review of its biology and management. Journal of Insect Science. 5: 40.

[5] Seewooruthun SI, Sookar P, Permalloo S, Joomaye A, Alleck M, Gungah B, Soonnoo AR. 1998. An attempt to the eradication of the Oriental fruit fly, Bactrocera dorsalis (Hendel) from Mauritius. In: Lalouette JA, Bachraz DY, Sukurdeep N, Seebaluck BD editors. Proceedings of the Second Annual Meeting of Agricultural Scientists, 12-13 August 1997, pp. 181-187. Food and Research Council, Reduit, Mauritius.

[6] Zaman M. 1995. Assessment of the male population of the fruit flies through kairomone baited traps and the association of the abundance levels with the environmental factors. Sarhad Journal of Agriculture 11: 657-670.

[7] Roomi MW, Abbas T, Shah AH, Robina S, Qureshi AA, Hussain SS, Nasir KA. 1993. Control of fruit flies (Dacus spp.) by attractants of plant origin. Anzeiger fur Schadlingskunde, Aflanzenschutz, Umwdtschutz 66: 155-157.

[8] Klungness LM, Jang EB, Mau RFL, Vargas RI, Sugano JS, Fujitani E. 2005. New approaches to sanitation in a cropping system susceptible to tephritid fruit flies (Diptera: Tephritidae) in Hawaii. Journal of Applied Science and Environmental Management 9: 5-15.

[9] Akhtaruzzaman M, Alam MZ, Ali-Sardar MM. 1999. Suppressing fruit fly infestation by bagging cucumber at different days after anthesis. Bangladesh Journal of Entomology 9: 103-112.

[10] Jaiswal JP, Gurung TB, Pandey RR. 1997. Findings of melon fruit fly control survey and its integrated management 1996/97, Kashi, Nepal. Lumle Agriculture Research Centre Working Paper 97/53, pp 1-12.0.

[11] Srinivasan K. 1994. Recent trends in insect pest management in vegetable crops. In: Dhaliwal GS, Arora R editors. Trends in Agricultural Insect Pest Management, pp. 345-372. Commonwealth Publishers, New Delhi, India.

[12] Wong TTY, Cunningham RT, Mcinnis DO, Gilmore JE. 1989. Seasonal distribution and abundance of Dacus cucurbitae (Diptera: Tephritidae) in Rota, Commonwealth of the Mariana Islands. Environmental Entomology 18: 1079-1082.

[13] Wood M. 2001. Forcing exotic, invasive insects into retreat: new IPM program targets Hawaii's fruit flies. Agricultural Research Washington 49: 11-13.

[14] Lindegren JE. 1990. Field suppression of three fruit fly species (Diptera: Tephritidae) with Steinernema carpocapsae. In: Proceedings 5th International Colloquium on Invertebrate Pathology and Microbial Control 20-24 August 1990, Adelaide, Australia, 1-223 pp.

[15] Sinha P. 1997. Effects of culture filtrates of fungi on mortality oflarvae of Dacus cucurbitae. Journal of Environmental Biology 18: 245-248.

[16] Sinha, P. and S.K. Saxena. 1998. Effects of culture filtrates of three fungi in different combinations on the development of Dacus cucurbitae in vitro. Indian Phytopathology. 51: 361-362.

[17] Sinha P, Saxena SK. 1999. Effect of culture filtrates of three fungi in different combinations on the development of the fruit fly, Dacus cucurbitae Coq. Annals of Plant Protection Service 7: 96-99.

[18] Nair S, Thomas J. 1999. Effect of Acorus calamus L. extracts on the longevity of Bactrocera cucurbitae Coq. Insect Environment 5: 27.

[19] Ranganath HR, Suryanarayana MA, Veenakumari K. 1997. Management of melon fly (Bactrocera (Zeugodacus) cucurbitae in cucurbits in South Andaman. Insect Environment 3: 32-33.

[20] Akhtaruzzaman M. Alam MZ, Ali-Sardar MM. Efficiency of different bait sprays for suppressing fruit fly on cucumber. Bulletin of the Institute of Tropical Agriculture, Kyushu University; 2000. 23:15-26.

[21] Gupta JN, Verma AN. 1982. Effectiveness of fenitrothion bait sprays against melon fruit fly, Dacus cucurbitae Coquillett in bitter gourd. Indian Journal of Agricultural Research 16: 41-46.

[22] Reddy AV. 1997. Evaluation of certain new insecticides against cucurbit fruit fly (Dacus cucurbitae Coq.) on bitter gourd. Annals of Agricultural Research 18: 252-254.

[23] Mishra PN, Singh MP. 1999. Studies on the ovicidal action of diflubenzuron on the eggs of Dacus (Bactrocera) cucurbitae Coq. damaging cucumber. Annals of Plant Protection Science 7: 94-96.

[24] Mumford JD. 2004. Economic analysis of area-wide fruit fly management. In: Barnes B, Addison M, editors. Proceedings of the 6th International Symposium on Fruit Flies of Economic Importance, Stellenbosch, South Africa, 6-10 May 2002. Infruitec Press, Stellenbosch, South Africa.

[25] Doharey, K.L. 1983. Bionomics of red pumpkin beetle, Aulacophora foveicollis (Lucas) on some fruits. Indian Journal of Entomology. 45: 406-413.

[26] Rahaman, M.A. and M.D.H. Prodhan. 2007. Effects of net barrier and synthetic pesticides on red pumpkin beetle and yield of cucumber. International Journal of Sustainable Crop Production. 2(3): 30-34.

[27] Rahaman, M.A., M.D.H. Prodhan, and A.K.M. Maula. 2008a. Effect of botanical and synthetic pesticides in controlling Epilachna beetle and the yield of bitter gourd. International Journal of Sustainable Crop Production. 3(5): 23-26

[28] Kamal, M.M., M.M. Uddin, M. Shajahan, M.M. Rahman, M.J. Alam, M.S. Islam, M.Y. Rafii, and M.A. Latif. 2014. Incidence and host preference of red pumpkin beetle, Aulacophora foveicollis (Lucas) on Cucurbitaceous vegetables. Life Science Journal. 11(7): 459-466.

[29] Khan, M.M.H., M.Z. Alam, M.M. Rahman, M.I. Miah, and M.M. Hossain. 2012. Influence of weather factors on the incidence and distribution of red pumpkin beetle infesting cucurbits. Bangladesh Journal of Agricultural Research. 37(2): 361-367.

[30] Vishwakarma, R., P. Chand, and S.S. Ghatak. 2011. Potential plant extracts and entomopathogenic Fungi against

Red pumpkin beetle, Raphidopalpa foveicollis (Lucas). Annals of Plant Protection Science. 19(1): 84-87.

[31] Khan, S.M. and M. Wasim. 2001. Assessment of different plant extracts for their repellency against red pumpkin beetle (Aulacophora foveicollis Lucas.) on muskmelon (Cucumis melo L.) crop. Journal of Biological Sciences. 1(4): 198-200.

[32] Ali, H., S. Ahmad, G. Hassan, A. Amin, and M. Naeem. 2011. Efficacy of different botanicals against red pumpkin beetle (Aulacophora foveicollis) in bitter gourd (Momordica charantia L.). Journal of Weed Science Research. 17(1): 65-71.

[33] Hasan, M.K., M.M. Uddin, and M.M. Haque. 2011. Efficacy of malathion for controlling red pumpkin beetle, Aulacophora foveicollis in cucurbitaceous vegetables. Progress Agriculture. 22(1 & 2): 11-18.

[34] Mahmood, T., M.S. Tariq, K.M. Khokar, Hidayatullah, and S.I. Hussain. 2010. Comparative effect of different plant extracts and insecticide application as dust to control the attack of red pumpkin beetle on cucumber. Pakistan Journal of Agricultural Research. 23(3-4): 196-199.

[35] Thapa, R.B. and F.P. Neupane. 1992. Incidence, host preference and control of the red pumpkin beetle, Aulacophora foveicollis (Lucas) (Coleoptera: Chrysomelidae) on Cucurbits. Journal of the Institute of Agriculture and Animal Science. 13: 71-77.

[36] Boucher, J. 2014. Squash beetle: *Epilachna borealis*. U. Conn. Extension. http://ipm.uconn.edu/documents/raw2/672/Squash%20beetle%20article.pdf.

[37] Mondal, S., and S.S. Ghatak. 2009. Bioefficacy of some indigenous plant extracts against epilachna beetle (Henosepilachna vigintioctopunctata Fabr.) infesting cucumber. Journal of Plant Protection Sciences. 1(1): 71-75.

[38] Ostermann, H. and M. Dreyer. 1995. The Neem tree *Azadirachta indica*, A. Juss. and other meliaceous plants sources of unique natural products for integrated pest management, industry and other purposes. In: Vegetables and Grain Legumes. Schmutterer, H. in collaboration with K.R.S. Ascher, M.B. Isman, M. Jacobson, C.M. Ketkar, W. Kraus, H. Rembolt, and R.C. Saxena (eds.). VCH Weinheim, Germany. pp. 392-403. ISBN: 3-527-30054-6

[39] Rahaman, M.A., M.D.H. Prodhan, and A.K.M. Maula. 2008b. Effect of botanical and synthetic pesticides in controlling Epilachna beetle and the yield of bitter gourd. International Journal of Sustainable Crop Production. 3(5): 23-26

[40] Swaminathan, R., S. Manjoo, and T. Hussain. 2010. Anti-feedant activity of some biopesticides on Henosepilachna vigintioctopunctata (Fab.) (Coleoptera: Coccinellidae). Journal of Biopesticides. 3(1): 77-80.

[41] Islam, K., M.S. Islam, and Z. Ferdousi. 2011. Control of Epilachna vigintioctopunctata Fab. (Coleoptera: Coccinellidae) using some indigenous plant extracts. Journal of Life and Earth Science. 6: 75-80.

[42] Capinera, J.L. 2005. Melon Worm. Diaphania hyalinata. Featured Creatures. Division of Plant Industry, Department of Entomology and Nematology Florida Cooperative Extension Service, Institute of Food and Agricultural Sciences, University of Florida, Gainesville, FL. Publication # EENY-163. http://edis.ifas.ufl.edu/in320.

[43] Barbercheck, M.E. 2014. Biology and management of aphids in organic cucurbit production systems. www.extension.org/pages/60000/

biology-and-management-of-aphids-in-organic-cucurbit-production-systems.

[44] Stapleton, J.J. and C.G. Summers. 2002. Reflective mulches for management of aphids and aphid-borne virus diseases in late-season cantaloupe (Cucumis melo L. var. cantalupensis). Crop Protection. 21: 891-898.

[45] Toba, H.H., A.N. Kishaba, G.W. Bohn, and H. Hield. 1977. Protecting muskmelon against aphid-borne viruses. Phytopathology. 67: 1418-1423.

[46] Koch, R.L. 2003. The multicolored Asian lady beetle, Harmonia axyridis: A review of its biology, uses in biological control, and non-target impacts. Journal of Insect Science. 3: 16-32.

[47] Laska, P., C. Perez-Banon, L. Mazanek, S. Rojo, G. Stahls, M.A. Marcos-Garcia, V. Bicik, and J. Dusek. 2006. Taxonomy of the genera Scaeva, Simosyrphus and Ischiodon (Diptera: Syrphidae): Descriptions of immature stages and status of taxa. European Journal of Entomology. 103: 637-655.

[48] Markkula, M., K. Tiitanen, M. Hamalainen, and A. Forsberg. 1979. The aphid midge Aphidoletes aphidimyza (Diptera: Cecidomyiidae) and its use in biological control of aphids. Annales Entomologici Fennici. 45(4): 89-98.

[49] Tomanovic, Z. and M. Brajkovic. 2001. Aphid parasitoids (Hymenoptera: Aphidiidae) of agroecosystems of the south part of the Pannonian area. Archives of Biological Sciences. 53: 57-64.

[50] Kos K, Tomanovič Ž, Petrovič-Obradovič O, Laznik Ž, Vidrih M, Trdan S. Aphids (Aphididae) and their parasitoids in selected vegetable ecosystems in Slovenia. Acta agriculturae Slovenica [Internet]. Biotechnical Faculty; 2008 Jan 1;91(1). Available from: http://dx.doi.org/10.2478/v10014-008-0002-9

[51] Hall, R.A. 1982. Control of whitefly, Trialeurodes vaporariorum and cotton aphid, Aphis gossypii in glasshouses by two isolates of the fungus, Verticillium lecanii. Annals of Applied Biology. 101(1): 1-11.

[52] Nyoike, T.W. 2007. Evaluation of living and synthetic mulches with and without a reduced-risk insecticide for suppression of whiteflies and aphids, and insect transmitted viral diseases in zucchini squash. A thesis submitted to the Graduate school in partial fulfillment for MS in integrated pest management. University of Florida, Gainesville, FL, p. 90.

[53] McAuslane, H.J. and H.A. Smith. 2000. Sweetpotato Whitefly B Biotype, Bemisia tabaci (Gennadius) (Insecta: Hemiptera: Aleyrodidae). EENY-129, University of Florida/IFAS Extension. University of Florida, Gainesville, FL. https://edis.ifas.ufl.edu/in286.

[54] Abrahamian, P.E. and Y. Abou-Jawdah. 2014. Whitefly-transmitted criniviruses of cucurbits: Current status and future prospects. Virus disease. 25(1): 26-38.

[55] Jelinek, S. 2010. Whitefly management in greenhouse vegetable crops. Primefact 1007, pp. 1-6. http://www.dpi.nsw.gov.au/_data/assets/pdf_file/0009/339588/Whitefly-management-in-greenhouse-vegetablecrops.pdf

[56] Horowitz, A.R. 1986. Population dynamics of Bemisia tabaci (Gennadius): With special emphasis on cotton fields. Agriculture, Ecosystems and Environment. 17: 37-47.

[57] Horowitz, A.R., H. Podoler, and D. Gerling. 1984. Life table analysis of the tobacco whitefly Bemisia tabaci (Gennadius) in cotton fields in Israel. Acta Ecological/Ecologia Applicata. 5: 221-233.

[58] Frank, D.L. and O.E. Liburd. 2005. Effects of living and synthetic mulch on the population dynamics of whiteflies and aphids, their associated natural enemies and insect-transmitted plant diseases in zucchini. Environmental Entomology. 34: 857-865.

[59] Summers CG, Stapleton JJ. "Use of UV Reflective Mulch to Delay the Colonization and Reduce the Severity of Bemisia Argentifolii (Homoptera: Aleyrodidae) Infestations in Cucurbits." Crop Protection 21, no. 10 (December 2002): 921-928. doi:10.1016/s0261-2194(02)00067-4.

[60] Webb, S.E., F. Akad, T.W. Nyoike, O.E. Liburd, and J.E. Polston. 2014. Whitefly-Transmitted Cucurbit Leaf Crumple Virus in Florida. IFAS Extension. http://edis.ifas.ufl.edu/pdffiles/IN/IN71600.pdf

[61] Cranshaw, W.S. 2014. Greenhousewhitefly.http://www.ext.colostate.edu/pubs/insect/05587.html

[62] Faria, M. and S.P. Wraight. 2001. Biological control of Bemisia tabaci with fungi. Crop Protection. 20: 767-778.

[63] Gerling, D., O. Alomar, and J. Arno. 2001. Biological control of Bemisia tabaci using predators and parasitoids. Crop Protection. 20: 779-799.

[64] Legg, J., D. Gerling, and P. Neuenschwander. 2003. Biological control of whiteflies in sub-Saharan Africa. In: Biological Control in IPM System in Africa. Neuenschwander, P. C. Borgemeister and J. Langewald, (eds.) International Institute of Tropical Agriculture, Benin Station, Cotonou, Benin.

[65] Snyder, W.E. and D.H. Wise. 2001. Contrasting trophic cascades generated by a community of generalist predators. Ecology. 82: 1571-1583. http://www.jstor.org/stable/2679801.

[66] Worthley, H.N. 1923. The squash bug in Massachusetts. Journal of Economic Entomology. 16:73-79.http://ww.ingentaconnect.com/content/esa/jee/1923/00000016/00000001/art00010.

[67] Pickett, C.H., S.E. Schoenig, and M.P. Hoffmann. 1996. Establishment of the squash bug parasitoid, Trichopoda pennipes Fabr. (Diptera: Tachinidae), in northern California. Pan-Pacific Entomologist. 72: 220-226.

[68] Palumbo, J.C., W.S. Fargo, R.C. Berberet, E.L. Bonjour, and G.W. Cuperus. 1993. Timing insecticide applications for squash bug management: Impact on squash bug abundance and summer squash yields. Southwest Entomology. 18: 101-111.

Chapter 11

The Activity of New Bio-Agent to Control *Cucumovirus Cucumber Mosaic Virus* (CMV)

Maadh Abdulwahab Alfahad

Abstract

CMV virus is worldwide, especially in temperate regions, where it can infect more than 800 plant species belonging to about 40 families. Although the main factor that the plant takes in order not to be infected is because it has preventive means that inhibit the direction of pathogens so that the infection occurs under conditions that suit it and suit its success. Cucumber Mosaic Virus belongs to the group of plant viruses to the genus Cucumovirus, as the virus particles are symmetrically spherical, not enveloped, with a diameter of 29 nm, and the virus has several strains that differ among themselves in terms of factors, symptoms of infection and methods of transmission. The stimulation of induced systemic resistance (ISR) leads to the interest of many researchers. Many types of research and studies have been conducted in the field of biochemical changes in the form of modulating the host's cell wall. The production of phytoalexin. And the manufacture of pathogen-related proteins (Pathogenesis Related Protein). It has been indicated that treatment with various factors, for example (non-pathogenic organisms, weak pathogens, chemical and industrial compounds, plant extracts, nutritional supplements) has the ability to activate plant defense mechanisms and induce systemic resistance against pathogens. In the field of biological control, bacterial types have been used on many pathogens, including fluorescens Pseudomonas and Bacillus subtillus, as they have proven effective in controlling many different fungal and bacterial pathogens as well as viral, and the reason is due to the ability of the bacteria to produce many growth regulators and thus stimulate resistance The systemic plant and the production of phytotoxins are in addition to being one of the most important growth stimuli. New methods have been used to resist viruses by using natural nutritional supplements with effective effect, because plants have defensive means, and for this reason, the use of these supplements can be stimulated in addition to the preventive aspect, a decrease in infection parameters, and an increase in growth indicators and outcome. Several methods have been relied upon to diagnose viruses, the first being the symptoms of reagents, and they are of basic methods. After that, serological tests were adopted, which are highly specialized and accurate in diagnosing viruses, and electron microscopy was used as a method to detect the size and shape of viruses. Polymerase Chain Reaction (PCR) technology is a fast and accurate way to detect plant viruses compared to other tests, such as the ELISA test and plant reagents.

Keywords: biological control, cucumber mosaic virus, spirulina, food supplements

1. Introduction

Plant diseases are one of the main problems facing agricultural production worldwide, as they cause great losses to basic food crops. They may also lead to a decline in agricultural production. It is one of the plants belonging to the family Cucurbitaceae, and it is one of the summer vegetable crops grown in Iraq and that it is available throughout the year through the use of the open and covered agricultural pattern, according to the conditions that suit production [1].

CMV virus is worldwide, especially in temperate regions [2], where it can infect more than 800 plant species [3] belonging to about 40 families [2]. Although the main factor that the plant takes in order not to be infected is because it has preventive means that inhibit the direction of pathogens so that the infection occurs under conditions that suit it and suit its success.

CMV belongs to the group of plant viruses to the genus Cucumovirus, as the virus particles are symmetrically spherical, not enveloped, with a diameter of 29 nm, and the virus has several strains that differ among themselves in terms of factors, symptoms of infection and methods of transmission [4].

The stimulation of induced systemic resistance (ISR) leads to the interest of many researchers. Many researches and studies have been conducted in the field of biochemical changes in the form of modulating the host's cell wall [5]. The production of phytoalexin [6]. And the manufacture of pathogen-related proteins (Pathogenesis Related Protein) [7]. It has been indicated that treatment with various factors, for example (non-pathogenic organisms, weak pathogens, chemical and industrial compounds, plant extracts, nutritional supplements) has the ability to activate plant defense mechanisms and induce systemic resistance against pathogens [8].

In the field of biological control, bacterial types have been used on many pathogens, including *Pseudomonas fluorescens* and *Bacillus subtillus*, as they have proven effective in controlling many different pathogens, and the reason is due to the ability of the bacteria to produce many growth regulators and thus stimulate resistance in plant and the production of phytotoxins are in addition to being one of the most important growth stimuli.

New methods have been used to resist viruses by using natural nutritional supplements with effective effect, because plants have defensive means, and for this reason, the use of these supplements can be stimulated in addition to the preventive aspect, a decrease in infection parameters, and an increase in growth indicators and outcome [9].

Several methods have been relied upon to diagnose viruses, the first being the symptoms of reagents, and they are of basic methods [10].

In this chapter of the book we will discuss the following topics:

1. Isolation and molecular diagnosis of local isolation of CMV virus by RT-PCR assay.

2. To test the sensitivity of three cucumber genotypes against CMV.

3. Evaluation of the natural food supplement produced from *Spirulina sp* against CMV.

4. Evaluating the effect of *P. fluorescens* inoculation on infection rate and severity of CMV infection and some growth parameters and yield of cucumber varieties against cucumber mosaic virus.

5. Evaluation of the biological efficacy of medicinal plant extracts (licorice, sespan, and albizia) in inhibiting CMV infection.

6. The effect of the complementary treatment with the treatments mentioned above on the seeds on the percentage of their transmission and germination strength.

7. Determining the rate of transmission of the virus in seeds from different sources (local and international).

2. Review of literatures

2.1 Cucumber mosaic virus

2.1.1 Symptoms

The pathological symptoms of *Cucumovirus Cucumber Mosaic Virus* are described, which are characterized by the yellowing and mottling that spreads on cucumber leaves and also on the fruits (**Figure 1**), which is a yellow mottling that affects all leaves and after infection the leaves are deformed in addition to stunting, which is accompanied by gradual plant degradation and curling of the leaves, as well as surface roughness. The outer layer of the leaf and the appearance of small prominent growths, these growths are clearly between the small veins [11]. Mosaic cucumber is from the Cucumovirus group that works on the appearance of mosaics and mottles on their families. Gibbs and Harrison [12] stated that this virus causes green mosaic or alternating green and yellow colors with each other on cucumbers.

The virus causes symptoms of bundling veins and deformation of the leaves, in addition to the appearance of mosaic and the curvature of modern leaves, and the necks of the leaves are reduced and accompanied by deformation of the fruits on

FIgure 1.
CMV pathological symptoms.

the pepper plant [13], but on the mosaic eggplant and dwarfism, it is clear, and necrosis rings sometimes appear on the leaves, general distortion and layout [14].

The virus generally causes clear symptoms of mosaic on the leaves and many fruit trees and other plants, in addition to general stunting of banana trees, ornamental plants, and gladiolus, where it produces symptoms similar to flower breakage. The virus has been diagnosed in Iraq and there are 4 strains that depend on the basis of serological and biotic properties as well as the migration of the protein coat. Another study showed that most strains of Group 1, A1 and B1 do not cause symptoms on the tobacco plant, but they infect species of Vigna spp. Systemically and some of them give localized stains on tobacco with a yellow color, while the strains of Group 11 have local and systemic stains on tobacco and are Very severe to mild leaf distortions follow depending on strain and environmental conditions [15].

2.1.2 Cucumber mosaic virus classification

CMV (CMV) is related to the genus Cucumovirus, group Cucumoviruses, family Bromoviridae and genus Cucumovirus [16–20].

2.1.3 The size and gravity of the virus

The virus particle is spherical, ie, Isohedral, with a diameter of 28–30 nm [21]. It refers to the group of viruses of similar dimensions, Isometeric Viruses.

The genome consists of three segments of the filamentous DNA that are single-stranded and the total number of nucleotides entering its composition reaches 8621 nucleotides [20]. The largest piece is (RNA-1) which has a molecular weight of 1.3x106 Dalton and consists of 3389 nucleotides, the second piece (RNA-2) has a molecular weight of 1.1x1060 Dalton and consists of 3035 nucleotides and the third piece (RNA-3) is smaller with a molecular weight of 0.8x106 Dalton It is composed of 2197 nucleotides [22].

Sill et al. [23] found that CMV particles were spherical in shape and 35 nm in diameter. Also Scott [24]; Smith [25] and Matheuos [26] have noted that they are spherical particles with a diameter of 28–30 nm.

Indicated that isometric particles that are polyhedral spherical (**Figure** 2) and have a diameter of about 30 nm [27, 28].

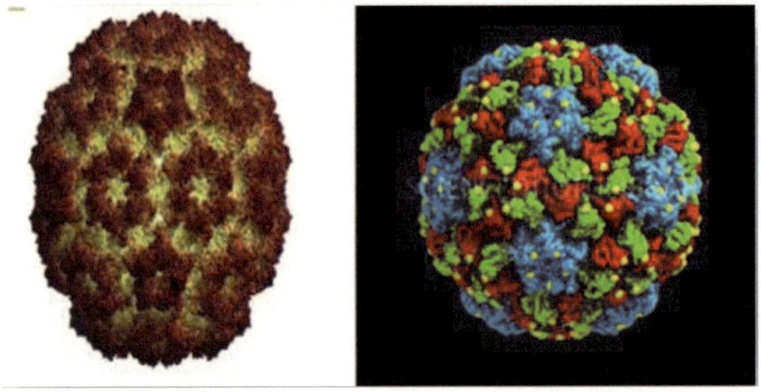

FIgure 2.
CMV shape under electron microscope.

2.1.4 Host range

It has been indicated by many researchers that the CMVhas a wide range of hosts. Also, Price [29] recorded the presence of 119 plant species of the virus, which belong to 40 families, and Nelson and Tuttle [30] mentioned that there are numbers of plants that exist. In Yama, Barisona, it belongs to many different families and is one of the important families of the virus, which are Pig weed, Wid beet, Nettlelleaf goosefoot, Lambis quarater, Saf flower, Winter synash, Periwinkle, Grownd cherry, Sowthistle.

CMV virus has the characteristic of spreading in all parts of the world because it has a wide host range and it works on infecting many types of ornamental trees and vegetables in addition to cucumbers [31].

2.1.5 Transportation methods

Insects are one of the most important means by which plant viral diseases spread in nature, and aphids have the main role among the insects that are capable of transmitting and spreading viruses, Maramorosch et al. [32] confirmed that the CMVis transmitted in an unsustainable manner by more than (60) types of aphids and that the transmission capacity varies according to the type of aphids, type of the host plant, and it is noted that the virus strain is more transmissible than other strains, and they also indicated The virus transmitted in this way will be completely lost within only two hours. El-Sayed et al. [33] found that it is easily transmitted CMV from the green peach *M. persicae* (**Figure 3A**) and *A. fabeae* (**FIgure 3B**).

This virus is transmitted by the seeds of the wild cucumbers that are adjacent to the cucumber fields, as well as the mulberry seed which is at a ratio of about (10–40%) in addition to that it is transmitted by the seeds of many jungles of annuals and is a small percentage. Transmission is accomplished by mechanical means, by means of contamination by hands, as well as easily by agricultural tools [34]. Franchi et al. [35] indicated that this virus is seed-transmitted, It can occur in 19 plant species, including some jungle plants. Mathews [26] has shown that this virus has the ability to survive in soil for a period of at least 12 years.

El-Sayed et al. [33] assert that CMV is not transmitted by the seeds of cultivated cucumbers, but that infection by viral infection leads to a reduction in seed weight by 23.5%.

2.1.6 The importance and spread of CMV

The CMV is one of the most important and widespread viruses, as it spreads in all regions of the world, and has a wide range of families, the most important of

(A) (B)

Figure 3.
Transmitted CMV by aphis. (A) Myzus persicae, *(B)* Aphis fabeae.

which are lettuce, celery, carrots, parsley, beets, walnuts, beets, spang, and many types of bush and ornamental plants [36]. It has been indicated that the CMV infects more than 1287 plant families, which belong to 518 plant genera and 100 species, CMV virus causes many economic losses to many crops [18, 37] and my understanding 2006).

It was recorded in America by Price [38]. In neighboring countries, it is considered one of the most important viruses that infect vegetable crops as well as ornamental plants, namely Jordan, Iran, Syria and Turkey [39].

2.2 Diagnosing CMV

Diagnostic methods have been used in a variety of ways to detect CMV virus by using a group of detector plants, which in turn give clear systemic symptoms, the most important of which are tomato *Lycopersicon esculentum*, watermelon *L. cucumis* melon, *Citrulli lanthatus* L. and cucumber L. Cucumber *sativus* and L. *Cucurbita pepo*, as well as there are plants that form symptoms in the form of local spots on arugula, barley and cowpea plants. This method has been adopted to diagnose the virus by many researchers.

There are many other methods through which the diagnosis is made by using anti-serum with many serological techniques that have been used to detect infected samples, the glass slide clumping method, double diffusion of clots, and ELISA tests were used [40].

Hu et al. [41] used the RT-PCR test, the reverse transcription-polymerase chain reaction, as well as the dot-blot assay to be detected, and the Tissue-Blot Immunoassay (TBIA) technology was also used [42].

It was also detected by cytofluorimetric fluorescence [43].

Immunostrip tests were used in the diagnosis of viruses and were successful, including CMV virus [44]. Use of protein electrophoresis on an acrylamide gel to detect the virus [45].

2.2.1 Polymerase chain reaction PCR technology

RT-PCR method is considered one of the most important rapid techniques and is effective and decisive in order to diagnose and determine the type of strain and know the difference between viruses that are similar and are used in disease control as well as control breading and mitigate the epidemic with all plant viruses [46]. A fast and sensitive method, which is inferred by detecting viruses in comparison to other tests.

This technique can be used to detect very low concentrations of viruses in plant tissue, and because of their accuracy as well as their sensitivity, it has been used to detect viruses and thus to identify strains successfully.

These specifications have led to applications in medicine, forensic analysis, agriculture, and mutation analysis in eukaryotic organisms. PCR technology depends on the replication of pieces of DNA by a type of DNA polymerase that is heat resistant outside the body of the in vitro to millions of copies by the presence of the primer. It is linked according to the sequence that completes it on the DNA Template with the presence of a sufficient amount of dNTPS, and the cloning and elongation reactions remain in a sequential manner, depending on the temperature change. As for the RNA, the previous steps of replication are tracked, which is converting it into cDNA by the method called RT-PCR. Same steps for replication and elongation of DNA.

2.3 The role of resistance with bio-agents and plant extracts in limiting the harm of viral diseases

2.3.1 Stimulating the systemic resistance of plants against plant pathogens by bio-agents

The plant can resist a "pathogen" or several pathogens through structural or biochemical defenses that help the plant to inhibit the pathogen. The plant defense mechanisms are a result of the activity of the gense resistance genes responsible for plant defenses [47].

Many different biological factors have been used in the resistance of pathogens, especially viral pathogens [48]. The first knowledge of induced resistance was recorded in 1933 by Chester, and the first to use this term was Ross in 1961 who inoculated the lower leaf of tobacco plants sensitive to (TMV) Tobacco Mosaic Virus and this induced resistance in the upper leaves against TMV [49].

Induced resistance is the resistance that is based on the structural and chemical defenses that are induced after inoculation with an unsatisfactory or pathogen incompatible with the plant host, so this type of resistance develops systemically as a response to the presence and settlement of PGPR bacteria. This resistance shows specialization in stimulating resistance.

Induced Systemic Resistance (ISR), unlike Systemic Acquired Resistance (SAR), is not related to the expression genes of the pathogen-related proteins of the acquired resistance. The induced resistance is either local or systemic and the meaning of resistance. Localization is the resistance that occurs by entering the pathogen into the tissues of the plant and thus results in the production of phytotoxins, as well as the grouping of two lignins so that they lead to strengthening the walls of the cells and difficult to penetrate by the pathogen or because it leads to the death and destruction of plant cells and their dehydration, which is called the hypersensitivity reaction (HR), GM plants in which the CP protein coat accumulates from the tobacco mosaic virus TMV are resistant to infection with TMV.

Resistance by the CP gene has been demonstrated against CMVCMV, clover mosaic virus AMV, potato virus PVX, tobacco streak virus, and other viruses [50].

Many theories have been developed to explain the stimulation of plant growth and resistance by these factors, and the most common is the secretion of antibiotics and the production of compounds that compete with chemical elements needed by the pathogen in its development, as well as stimulating resistance genes with these materials and editing the work of the operator genes by disengaging them from the repressor protein molecule. It is reflected in the manufacture of anti-virus materials that may be proteins, including enzymes linked to the virus and that prevent the release of DNA [51, 52].

It was found that *Pseudomonas fluorescens* and *Rhizobacteria leguminosarum* induced systemic resistance in plants against BYMV [53] tested the possibility of stimulating pepper plant resistance against PMMOV.

Defensive gene products that include peroxidase enzyme (po) and polyphenol oxidase (ppo), which are concerned with the combination of lignin and phenaline ammonialase (pal) as they have an important role in the manufacture of phenols, phytotoxins and other defense enzymes that include pathogen-related proteins (pr), for example b1,3_glucanases (pr2 family), chitinase and the lipase enzyme (pr3 family), which causes the decomposition of the fungal cell walls and thus the complete analysis of the cell when a fungal infection occurs that works to dissolve the lipoprotein envelope of the virus if the infection is an enveloped virus [54].

Several researchers have discussed the possibility of inducing cucumber and tomato plants to be resistant to CMV(CMV) as well as tomato mottle virus (TOMOV) using isolates from *P. fluorescence, P. putida* and *Bacillus pumils* [52]. The application of *P. fluorescence* induced SR in cucumber against CMV.

2.3.2 Pseudomonas fluorescens

P. fluorescens bacteria represent a large and important group of Gram-negative bacteria that are present in large numbers as well as are free to live in soil, fresh water and salty environments, especially in the vicinity of the roots due to their ability to grow on organic materials that are abundant around the roots and many other natural environments,

It has received great attention within the term Induced Systemic Resistance (ISR) due to its effect on protecting plants treated with it as it is considered one of the most important types of root bacteria that stimulate plant growth [55], and in addition it has beneficial effects to promote plant growth. And that by providing the treated plants with many growth regulators, including gibberellins, auxins, and others [56].

Al-Fahad [57] has indicated that *P. fluorescens*, when used as a bacterial vaccine, has the effect of improving or increasing and stimulating growth, which is represented by increasing the leaf area, chlorophyll, in addition to stimulating the resistance of plants to infection with the virus or reducing its multiplication within plant tissues, and this positive increase was significantly and clearly reflected on the outcome.

Al-Fahad [52] mentioned that there are several theories that explain this stimulation, including the secretion of antibiotics by the stimulus factor outside or inside the plant. P.F bacteria were used in plant resistance as inducers against CMV, and it led to a decrease in the severity and rate of infection with the virus.

It grows at an optimum temperature of 20–30° C and most strains grow at 4° C [58]. Kazenpour [59] found that treatment in P.F with rice seed reduced the severity of rice sheath blight by 10.5%, while in comparison it was 52%.

2.3.3 Spirulina platensis

We have been heading to thinking about resistance to viruses by stimulating resistance against it, due to the difficulty of controlling it by traditional methods. One way to enhance plant resistance against viruses is the use of the algae *S. platensis*.

It was recently classified by Ali and Saleh [60] and Kewi et al. [61] as a type of bacteria, **Figure 4**.

Kingdam (Domain): Eubacteria, Bacteri

Figure 4.
S. platensis *form under a microscope.*

Phylum: Cyanobacteria
Order: Spirulinalies
Family: Spirulinaceae
Genus: *Spirulina*
Species: *Platensis*

Algae are ancient discovered organisms that date back to fossils about 305 billion years ago [62]. Its importance for health and the prevention of various diseases has been mentioned [63].

Spirulina is a bluish green algae that has a filamentous and spiral shape and its taxonomic position is unique because it combines Autorrphic [64], which is common in eukaryotic cells as well as bacteria and has high nutritional value, for example, vitamins B1, B2, and A [65]. The chemical composition is from proteins 55–70%, carbohydrates 15–25%, essential fatty acids 18%, minerals, vitamins, dyes, for example carotenoids and chlorophyll [60]. It was recently classified as one of the rare edible bacteria due to its low concentration of purine [66]. *S. platensis* is widely used as a nutritional supplement for human health, as well as animal feed, and its importance lies due to its high protein content of 60–70% and its high concentration of essential amino acids, fatty acids, minerals, vitamins and antioxidants [66, 67].

Blinkova et al. [68] showed that Sulfolipid extracted from *S. platensis* inhibits the activity of HIV and also consists of many amino acids and sugars and is considered essential compounds in addition to the micro and macro elements that have the essential role to improve plant growth [69].

Blue-green algae are highly efficient in producing a wide range of antibiotics that have a direct effect in inhibiting the growth of pathogenic bacteria. The anti-bacterial effect of marine algae is not limited to bacteria but also has an anti-viral effect [70]. Green algae have a high efficiency to produce a group of antibiotics and have a direct effect in inhibiting the growth of pathogenic bacteria P.sp., which is resistant to antibiotics. It is also considered one of the most important organisms due to its wide range and bioactivity including: This effect on viruses is inhibitory [71].

The results obtained by Buter and Hunter [72] have supported the addition of extracts consisting of seaweed to improve the growth and productivity of the plant, and that the process of foliar spraying through these extracts has led to an increase in vegetative and productive growth and the reason is to provide essential nutrients that the root cannot provide. Mishima [73] showed that *S. platensis* has the ability to inhibit the multiplication of many viruses.

Spenille and others [74] indicated that algae have significant effects as they increase the resistance of plants against disease, when added to the plant, and it is one of the most important organic sources used in plant production. The anti-effect of marine algae does not depend on the bacterium only, as it has an anti-viral effect [70]. Algae are important biological groups due to their wide range and biological effect, including the inhibitory effect against viruses [71]. The treatment of pepper plants with *S. platensis* had a positive effect in reducing the severity of ToMV mosaic virus, by reducing the phenotypic symptoms, the activity of the peroxidase enzyme, and the increase in leaf area, plant height, chlorophyll percentage, and dry weight of the shoots [75]. The United Nations Food and Agriculture Organization (FAO) as well as the independent governmental organization, www.iimsam.erg [76], recommended and emphasized the necessity of using SP as a basic and main tool to combat malnutrition in the world as well as to help achieve sustainable development. In India, some villages were working on the use of blue-green algae to improve the quality of saline soil in Samphar in Rajasthan, it is dried and used as fertilizer for crops such as wheat and barley. Gupta et al. [77] used some types of blue-green algae (*Arthrospira subsalsa & Spirulina Platensis*) with bio-fertilization and mitigation of salinity damage.

2.3.4 Medicinal plant extracts

Alternatives to chemical pesticides such as plant extracts have been used to protect plant production from diseases caused by fungal, bacterial, viral and nematode organisms, as they cause great losses to the economy during the growing season or in the post-harvest stages.

It was found that the extracts of pepper, geranium and datura plants have the ability to induce resistance to viral infection in the paper that was treated before inoculation with the virus, as well as act as infection inhibitors when mixed with the virus, and it was found that they contain compounds of partial weights, low and high that have effectiveness against viral infection [78].

2.3.4.1 Sesabania spp.

The *Sesabania spp*. Belongs to the family of the leguminous Fabaceae or Leguminosae, and the leaf is a main compound, the stamens have many flowers and the fruit is a pod and contain a group of seeds. Therefore, many countries paid attention to this plant. Ndungiu [79] mentioned that there are many international institutions in East African countries that they have conducted a complete and comprehensive survey of the varieties that spread in countries, Malawi, Zambia, Botswana, Namibia, Zimbabwe.

Sesbania sesban Linn otherwise known as "Sesban" (**Figure 5**) is one of the six species of the genus Sesbania that is commonly found for cultivation in the tropical region of India. The plant is widely cultivated for its ability to stabilize nitrogen and wind shade, obtained good medicinal importance according to ethnomedical claims [80].

The plant contains many phytochemicals, including saponins terpenoids, which are fat compounds that are composed of the bindings of B-amyrin, Licoric acid and Glabrolide, Among them is Isoprene, Osiprene called Penta-hydrocarbon, Glycyrrhetol, Glycyrrhizin and Liguiritic acid. The effective natural products present in the plant S. sesban as a means of finding alternatives to the chemical pesticides manufactured because of the effective compounds this plant contains, which include phenolic compounds, including saponins [81].The leaves of the plant contain many important components that are found within the chemical components of the plant, including Alkaloidl, Flavonid, Tannius, Terpenad.

The use of this plant to get rid of pathogens by using it in the form of an alcoholic and aqueous extract or in the form of powder, and it also has high efficacy

Figure 5.
Sesbania sesban *plants.*

against fungi and bacteria and that these are isolated from Sespan, which have the effect of flavonoids [82, 83].

2.3.4.2 Albizia spp.

Albizia is widely spread in the world and the scientific name of Albizia is (L) Albizzia lebbeck and belongs to the family of Leguminosae. Its leaves are large and a fragrant group of yellow flowers and long seeds. It belongs to the family Leguminosae and is found in tropical Asia and is widely cultivated in other tropical and subtropical regions including Malawi [84]. Its flowers, bark, roots and stems are all used in medicine [85].

Albizia is an important plant as an important source of pesticides and because it contains important chemical compounds such as Saponins Alkaloids, Flavonids, Tannius [86]. The common name is bacha chin, ficus and acacia, and it is widely cultivated in the regions of tropical Asia, Africa and subtropical Asia [87].

It has many uses, including medicinal, environmental and economical, and is used as shade trees as well as wood production, and is considered to be of natural importance [88]. Its seeds contain lectins that are toxic to humans and animals.

2.3.4.3 Glycyrrhiza spp.

The (Licorice) *Glycyrrhiza glabral,* which belongs to the Leguminosae family, and one of the most famous genera is the Glycyrrhizin plant, which contains 20 species. Glycyrrhiza, which means sweet veins in Greek.

Licorice contains many chemical compounds, as it is characterized by the availability of chemical compounds that have a sweet taste, and it contains Glycyrrheiel, Liguoric acid and flavonoids, including Glabridin and Glabrin [77]. Glycyrrhizin with its acid is the most important ingredient, it has an activity similar to that of steroid hormones, i.e. plant hormones are formed and thus lead to an increase in protein formation and an improvement in the indicators of vegetative growth [89].

The plant spreads in Asia Minor and Turkmenistan in the east to Spain in the west, which are its original home and also because of its medicinal value It has become cultivated on large areas in France, Spain, Germany, Russia, Italy, China, Turkey and America as well as the Arab Gulf and Iraq [90].

Al-Ajili [91] emphasized that licorice has high concentrations of amino acids as well as carbohydrates and other important elements, which leads to its great importance in nutritional uses. Licorice, G. glabral, a type of the Leguminosae family, is of great medical importance, as it is used for treating stomach irritation, coughing, hoarseness, bladder and kidney inflammation, as well as gout and muscular rheumatism, the autumn season are among the best dates for collecting licorice, and the reason is because the active substances are more abundant in this season, and Autumn is the best time to collect licorice because the active ingredients are more abundant [92].

Chemical analysis of licorice extract revealed that there are phenolic compounds, terpenes, gummies, starch, and flavonids extracts due to phenolic compounds for resistance to the microbes *Candida albicans*, *Staphylococcus aureus*, and Mycobacterium semegmutis, and the effect of extracts of ten species, including licorice, which showed very high efficacy in inhibiting TYLCV, reaching 100% for all concentrations used [93].

Licorice leaves are used to extract antimicrobial agents such as Pincembrin and Licoflavanone [94]. And the extract is used to protect against atomic radiation, as there is in the extract the substance alpha-amyrin that combines with the substance

Methyluracil to form a compound that works to protect against atomic radiation, Maliuta [95]. And the World Health Organization (WHO) stated that licorice extract treats laryngitis and is used as an expectorant to treat colds and coughs, as well as bronchitis, to treat rheumatism, to protect the liver from poisoning and to arthritis, and to treat tuberculosis [96]. The active ingredients are Glycyrrhizin, Flavon Glycosides, Liquiritoside,Isoliquiritoside, Sucrose, Dextrose, Starch, Proteins, Liposomes, Resin, Asparagine, Sterilic Oil, Saponins [97].

3. Practical experiments followed in diagnosing and controlling cucumber mosaic virus

3.1 Collection of infected plant samples

As provided in [75].

3.2 Laboratory experiments

As mentioned in [52] with some modifications.

3.2.1 CMV molecular diagnostics

The local isolate mentioned in paragraph 3.1 obtained by Prof. Dr. Maadh AbdulWahab Al-Fahad was molecularly diagnosed by means of the technique of polymerase chain reaction, RT-PCR and according to the following steps, which were obtained from the producing company (BiONEER) for the kit (Diagnostic kit) with some modifications which included:

1. Isolation of RNA

2. Solutions used to isolate RNA

3. Isolation of Total Viral RNA

4. Measurement of extracted RNA purity and concentration

5. DNA replication by using RT-PCR technique

3.2.2 The solutions used in the migration process

A. SB solution with a strength of 1X.

B. B-Loading buffer 10x magnification.

Preparation of the Agarose Gel and Electrophoresis Procedure:

3.3 Prepare spirulina, bacteria and plant extracts

3.3.1 Spirulina

The nutritional supplement represented by *Spirulina platensis* was obtained from the Malaysian company DXN, as 24 spirulina tablets were taken, equivalent to 6 g of spirulina, crushed and placed in 54 ml of distilled water, each 1 g, 9 ml of distilled water was placed [75].

3.3.2 PF bacteria

An isolate of *Pseudomonas florescence* (pf) was obtained from the biocide Bactvipe provided by International Panacea Ltd. Sterilized and ready, the hood was sterilized and the medium was incubated in the incubator for a period of (72 hours) at a temperature (38° C). It was observed that the growth rate of bacterial colonies reached 100% of the dishes used in agriculture.

The dilution was used (10–4.10-5.10-6.10-7) and the fastest growing dilution and the best was the 6–10 dilution of the biological product where the plate was taken and added to a baker containing 500 ml distilled water to soak the seeds in it for 2 hours.

3.3.3 Extracts (sespan, albizia, and licorice)

A water-based extraction method was used to obtain the active compounds in the licorice plant, and this method was done according to Al-Ajili [91].

3.3.4 CMV virus extract

The method mentioned by Qasim and Ali [98].

3.4 Determination of amino acids of plant extracts

Amino acids were estimated based on the method presented by Scriver et al. [99] and according to the conditions attached in table below:

Time	Methanol A	Buffer B	Flow rate
09	20	80	1 ml/min
10–13	40	60	1 ml/min
14–25	90	10	1 ml/min

3.5 Determination of active compounds in the dry leaves of medicinal plant extracts

The active compounds were diagnosed in the dry leaves of Sispan, Albizia and licorice roots by using the High Performance Liquid Chromatography (HPLC) device. The dry leaves of Albizia, Sisban and licorice roots were extracted according to the method [100, 101]. The concentration of the active compounds was measured using the following model concentration equation:

$$\text{Sample concentration mg/ml} = \frac{\text{Space package form}}{\text{Area of the standard solution}} \times \text{Standard solution concentration} \times \text{Number of dilutions} \quad (1)$$

3.6 Field experiment

3.6.1 Plant varieties used in the experiment

The plant varieties referred to in the **Table 1** below and the companies producing them were obtained from the local markets for the sale of agricultural supplies.

3.6.2 Methods of treatment with materials used in the experiment

A. Albizia: The seeds prepared in the experiment were treated according to the scheme, by drenching (20) seeds for each line from the extract, for a period of (2) hours, and then sown in the field directly according to the scheme prepared for the experiment.

As for the second treatment, it was sprayed only on the third and fourth real leaf, and after grinding the leaves of Albizia (5) grams per liter of distilled water only. The third treatment was at the beginning of flowering in the same way as the second treatment.

B. Licorice: Licorice was treated in the same way as mentioned in Paragraph A, but for the roots.

C. Sespan: I prepared the treatment of the extract of sespan leaves in the same manner mentioned in Paragraph A.

D. Spirulina: The moss suspension was prepared by adding 1 g of *S. platensis* powder to 9 ml of sterile distilled water and the volume was completed to 1 liter of distilled water and soaking 20 seeds in it for 24 hours. The second treatment was spraying at the flowering stage. The third treatment was at the beginning of flowering in the same way as the second treatment.

E. P.F bacteria: The seeds that were prepared for cultivation were treated according to the scheme of the experiment, by soaking (20) seeds in the concentration (106) prepared in advance in the laboratory experiment, for a period of 24 hours. The second treatment was spraying at the flowering stage. The third treatment was at the beginning of flowering in the same way as the second treatment.

3.7 Measuring indicators of CMV infection

3.7.1 Calculate CMV infection ratio

It has been calculated per treatment according to the following equation:

$$\text{Infestation\%} = (\text{Infected plants number})/(\text{Total number of plants}) \text{ X } 100 \quad (2)$$

NO.	Name of varieties	Type of varieties	Producing company	Origin	Symbol
1	Superina	Hybrid	Nickerson-Zwaan	Holland	A
2	NAJIM	Hybrid	Seminis vegetable seeds	Chile	N
3	GHAZEER	Hybrid	Seminis vegetable seeds	USA	G

Table 1.
The plant varieties used in the experiment.

3.7.2 Measuring the severity of CMV infection

A guide (**Table 2**) to the severity of the infection was made by Prof. Maadh Abdel Wahab Al-Fahad, consisting of six degrees to know the development of the apparent symptoms on the plants. It was calculated by the number of affected plants in terms of the degree of each plant [102].

$$\text{Severity of infestation\%} = (0 \times 0 \text{ grade of plants number} + \ldots + 5 \times 5 \text{ grade of plants number}) / (5 \times \text{examined for plants total number}) \times 100 \quad (3)$$

Photo	Description	The degree of injury
	Healthy	0
	Light mosaic	1
	Severe mosaics on the leaves	2
	Severe mosaic and beginning wrap the leaves	3
	Intense yellowing, mosaic and severe leaf curl	4
	Severe mosaic and leaf wrap With deformations and shortening of the leaf space	5

Table 2.
Disease index of severity of CMV infection, by Dr. Maadh Alfahad.

The experiment measurements also included the following characteristics: Measuring the amount of chlorophyll, measuring the leaves area, measuring growth and outcome indicators, estimate the amount of total yield/plant, measure the plant height (cm. plant), measurement of dry weight of shoots/g, measurement of the dry weight of the root mass.

3.8 Statistical analysis

The experiment was carried out according to the independent randomized complete block design (RCBD) and the averages were compared according to the Least Significant Defense (LSD) test at a probability level of 0.05.

4. Results and discussion

4.1 Isolation and diagnosis of cucumber CMV mosaic virus

4.1.1 Symptoms

The symptoms appeared in mosaic on cucumber plants inoculated with CMV virus after a week of inoculation and over time the symptoms turned into deformation of the leaves and their edges bending down and twisting and stunting of the plant, as well as shortening of the necks of the leaves in addition to the lack of fruit numbers, deformation and dropping of flowers, and with the progression of the infection Areas wherein turns into yellow areas and then into dead areas (Necrosis), and no symptoms of infection with the virus appeared on healthy plants protected from infection.

4.1.2 CMV molecular diagnostics

4.1.2.1 Results of RNA isolation

Isolation of DNA from the leaves of cucumber plants infected with the virus, and an appropriate percentage of RNA was obtained, as the Kit (ready-made kit) method was used to obtain RNA in appropriate quantities ranging between (12.696–7.420) nanograms per microliter and the purity was (1.71) This result came after repeated attempts because of what is known that the process of isolating RNA is more difficult than DNA because it is sensitive to cracking by the RNase enzymes (Hassan, 2004).

4.1.2.2 Results of PCR reactions

A. RNA concentration.

B. The concentration of the initiator that was used.

C. Set the appropriate program to operate the thermal polymerization device.

When using the program mentioned in paragraph (3.3.1.5) the straight segments responsible for forming the virus' protein envelope were enlarged with the fourth primer (Forward 4: H), where the reaction showed very clear Bands with size (207 bp), and as They are shown in **Figure 6**. As for the primers Forward 1,

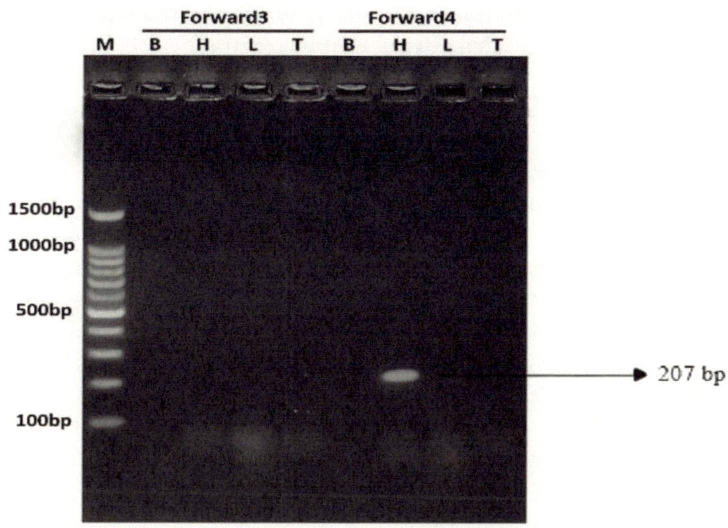

Figure 6.
A result of molecular detection of the local isolation of CMV virus for the third and fourth initiator. M: represents the volumetric index. Forward 4: H: represents the initiator of 207 bp for the fourth primer. B, L, T: Forward 3 represents the third primer. No package appeared.

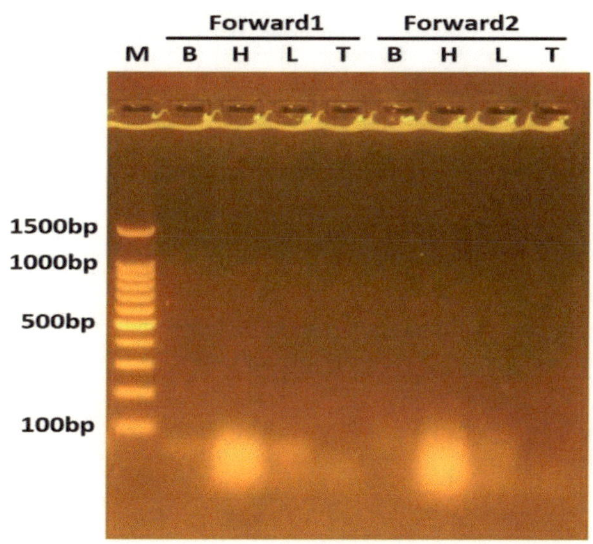

Figure 7.
The result of molecular detection of the local isolate of CMV virus for the first and second initiator. M: represents the volumetric index; Forward 1,2: represents the first and second primers where no bundles appeared; H: stands for cucumber. B: represents the pepper. L: represents tomato; T: symbolizes Al-Taroozi plant.

Forward 2 and Forward 3, they did not show any inflation as in **Figure 7**. This result of the molecular diagnostic test is similar to the findings of Shomaila Igbal et al. [103] in Rawalpindi region/ State of Pakistan, while the rest of the primers did not match, and this indicates a great convergence between the aforementioned isolation in Al-Huwaish/ Samarra district/Salahuddin with the Pakistani isolation. And a difference from the isolation in the regions (Punjab, Sind, Northwest Frontier Province, Islamabad and Baluchistan).

4.2 Evaluation of the effectiveness of some biological agents in inducing plant resistance against CMV virus in cucumber plants

4.2.1 The effect of the treatments used on the percentage of infestation to plants

The reason for the superior treatment of spirulina, P.F, and licorice (**Table 3**) may be attributed to the fact that spirulina algae produces biological compounds that have a toxic and anti-biotic effect for a group of microorganisms, including viruses.

Also, the low rate of infection of cucumber plants with CMV virus is due to the role of P.F. as a catalyst for systemic resistance against viruses by treating the seeds with the bacteria before planting. The reason for this may also be due to the fact that these bacteria induce resistance in plants against the CMV virus, and it led to a decrease in the rate of infection with the virus.

Treatments	Varieties							
	A	BEIT ALPHA	G	GHAZEER	N	NAJIM	Impact rate of treatments	
A	d-k	46.67	b-j	57.33	d-k	44.67	d-e	49.55
B	c-k	51.33	b-j	59.00	d-k	46.33	c-e	52.22
C	c-k	52.67	b-j	59.00	b-j	59.67	b-e	57.11
P.F	h-k	37.33	b-f	66.00	d-k	44.33	d-e	49.22
S	b-e	67.33	f-k	40.33	c-k	52.00	c-e	53.22
A+B	b-j	60.33	b-d	69.00	a-c	76.67	b	68.66
A+C	b-j	58.00	b-f	66.33	b-j	59.67	b-d	61.33
A+P.F	c-k	52.67	b-j	59.00	b-e	67.33	b-e	59.66
A+S	c-k	52.00	a-b	81.00	b-j	63.00	b-c	65.33
B+C	b-f	64.33	b-k	54.67	b-e	67.00	b-d	62.00
B+P.F	j-k	36.67	b-f	63.67	b-j	59.33	c-e	53.22
B+S	i-k	37.00	c-k	51.33	b-d	68.33	c-e	52.22
C+P.F	b-k	55.33	b-k	54.67	a-c	76.00	b-d	62.00
C+S	d-k	48.67	b-f	65.67	b-j	62.33	b-e	58.88
P.F+S	b-f	66.67	k	30.00	b-j	61.67	c-e	52.77
A+B+C	c-j	60.67	d-k	48.00	b-d	70.67	b-e	59.77
A+B+P.F	b-g	64.00	b-f	65.33	c-k	50.67	b-e	60.00
A+B+S	b-f	65.67	d-k	49.00	b-j	59.00	b-e	57.88
B+C+P.F	b-i	63.33	c-k	52.00	b-f	66.67	b-e	60.66
B+C+S	b-j	63.00	d-k	47.67	d-k	47.00	c-e	52.55
B+P.F+S	e-k	41.00	b-j	59.00	g-k	37.67	e	45.88
CO1	0.00	l	l	0.00	0.00	l	f	0.00
CO2	a	100.00	100.00	a	100.00	a	a	100.00
impact rate of var.	54.11	a	56.43	a	58.26	a		

**According to the Duncan polynomial test at the level of 5%.*

Table 3.
*The effect of the treatments used in the incidence of CMV infection % on three varieties of cucumber.**

The explanation of the superiority of the triple treatment between *S. platensis*, licorice and p.f bacteria may be attributed to its containment of most of the compounds needed by the plant to improve growth and also increase the efficiency of the photosynthesis process, and this is reflected in the inhibition of viral infection.

Or perhaps the reason for this is that *S. platensis* has inhibitory activity, which is clearly because it contains polysaccharides, cyclic peptides and alkaloids.

Al-Fahad [57], citing Jeanneus and Tetau [104], also mentioned that licorice extract contains Glycyrrhizic acid, which has medicinal efficacy and has affected the inhibition of the virus by forming hydrogen bonds linked to the virus protein or DNA or both, and this led to the limit of its effectiveness. Studies have indicated the possibility of stimulating plants to produce virus-inhibiting compounds as mentioned by Maurhafer et al. [105]. Or perhaps the reason for this is due to the fact that the amino acid Thereonin present in licorice extract led to an increase in the inhibition of the virus and reduced the rate of infection, as shown in **Table 3** and also the active substance that was disclosed in **Table 4** and **Figure 9**.

4.2.2 The effect of treatments on the severity of infestation to plants

The reason may be that the compound glycirizine with its acid present in licorice, which has an activity similar to that of steroid hormones, i.e. it is a form of plant hormones and leads to an increase in the formation of proteins, as shown by Tyler [89]. Also, Al-Janabi [93] concluded that licorice, which showed a very high efficacy in inhibiting TYLCV, reached 100% and protected tobacco plants against infection with TMV virus for 30 days.

This may be due to the fact that licorice extract contains terpenes, phenolics and starch, as well as resins, and flavonoids are due to phenolic compounds that resist many microbes.

The superiority of the interaction treatment between *S. platensis* and PF bacteria can be explained by the fact that the algae possesses materials rich in proteins, vitamins and minerals in addition to fatty acids, polyphenols and sugars and contains pigments such as carotenoids and chlorophyll that stimulate the growth and resistance of plants to pathogens and this is consistent with what he referred to. Abbasssy et al. [106] and Usharani et al. [107]. Also, the results of this study were consistent with the findings of Kim [55] that PF bacteria induce systemic resistance (ISR) of plants against diseases, as the study agrees with what Murphy and others [108] and Ryu and others [109] and EL-Dougdoug et al. [110] have indicated, with different mechanisms, including that the PF bacteria that are isolated from the roots are of great importance as growth stimuli and are considered as biological control agents for plant pathogens. These bacteria were used to induce resistance to plants against viruses or the secretion of multiple enzymes that help in analyzing the

No.	Plant name	Compound name	Concentration
1	Albesia	Keamferol	21.61
2	Licorice	Apigenen	76.053
3	Sespan	Gallic acid	1.196
		Qurcetine	3.402
		Ellagic acid	6.181

Table 4.
The effective compounds of medicinal plant extracts.

protein coat of the virus and also help produce Phytoluxins, which in turn are defensive compounds within plants [111].

4.2.3 The effect of used treatments on ratio of chlorophyll \ Spad, Leaf area/cm^2, plant length/cm and Yield amount/g, of three varieties of cucumber infected with CMV virus

The reason for the superiority of the treatment of spirulina, PF bacteria and licorice root extract may be attributed to the fact that the treatment with bacteria has a great role in stimulating the systemic resistance of plants and thus reduces the rates of effect of the virus on the amount of chlorophyll. The varieties, and also the tolerance to infection with the virus, may explain this effect that the bacteria PF leads to an improvement and increase the stimulation of growth, which is reflected in the increase in the amount of chlorophyll positively by increasing the chlorophyll.

Or, perhaps the reason for this is that it has the ability to form proteins that inhibit the replication of the virus by stimulating resistance genes in the plant that are associated with the virus and prevent the release of DNA, and thus the replication of the virus fails and is positively reflected in the amount of chlorophyll.

Several studies have shown that *S. platensis* contains growth regulators, the most important of which are cytokines that contribute to increasing chlorophyll synthesis, as well as compensating for the imbalance of nitrogenous bases as a result of infection with the virus. It was referred to by Mao and other [112].

The reason for the superiority of the variety *S. platensis* can be attributed to its possession of genes that carry resistance to the effects of the virus from breaking down chloroplasts, and is consistent with what was mentioned by Al-Fahad [57].

The difference in the chlorophyll percentage may be due to the genetic factor of the genes for resistance, because whenever they excel, they increase the percentage of resistance and thus reduce infection or reduce the (Virions) necessary for the formation of infection and lead to an increase in the chlorophyll percentage of the plant, and this is in line with what was mentioned by M. Others (2011).

This may be due to the fact that PGPR bacteria prevent the harmful effects of many different pathogens such as bacteria, fungi, nematodes, viruses, and produce materials that inhibit the growth of pathogens and have no harm to plants by providing the iron element that is necessary for the growth of these pathogens by Sidrophores and production Antibiotics [113].

Also, treating the seeds with soaking during planting increases the continuity of growth despite their injury, which makes the plant tolerant of the primary infection, and this is consistent with what Hassan and Jumaa [114] found that the significant increase in the amount of chlorophyll is caused by the soaking and spraying of growth stimuli.

Al-Ani [115] also indicated that licorice roots are used in the form of vegetable fertilizer because they contain nitrogen, so this may have been directly reflected in the increase in the amount of total chlorophyll, as nitrogen is of great importance in plants and this importance is through its entry into the building of many compounds necessary for growth And the continued growth of the plant, and introduces the building of photosynthesis pigments and the formation of energy compounds (NADH2, NADPH2, ATP) and bermidine and purine bases for nucleic acids and the formation of cell membranes, chloroplasts and mitochondria [116] (**Figure 9**).

Since it focuses on nucleic acids, it may contribute to compensating the plant for what was damaged by the presence of the virus and used the DNA for its benefit, and the reason may be that the amino acid Thereonin present in licorice extract led to an increase in the inhibition of the virus, unlike other compounds that were discovered and were in lower proportions From thereonin, this was reflected in the increase in chlorophyll because the plant assimilates it and uses it in protein syntheses and helps to increase the concentration of chlorophyll and achieve the highest level of photosynthesis and thus increase the vegetative growth and the amino acids are chelating substances that increase or improve the transport of nutrients [117] as shown in **Figure 9** as well as the active substance Apigenen, which was detected with licorice extract, has the greatest effect in increasing chlorophyll.

Perhaps the reason for the superior treatment of algae, bacteria and licorice is due to the fact that *S. platensis* contains phosphorous, which contributes to an increase in the vegetative system as well as the leaf area and is reflected in the plant's ability to absorb nutrients and this is positively due to the increase in the products of photosynthesis for the manufacture of foodstuffs.

The licorice extract is rapidly penetrating inside the leaf tissue cells, thus causing an effect on plant sensitivity or by influencing the virus replication process.

The reason for the plant height may be attributed to the algae *S. platensis* containing the nutrients that are sufficient for the seedlings to grow in the first stage in a strong way and thus bear the infection, because it contains many growth-promoting substances such as amino acids, organic acids, oxins, vitamins and cytokines, which have the most important role in the growth process and activating enzymes. It stimulates plant growth and also stimulates cell division and achieves the process of escaping from disease, and this is consistent with what was mentioned by Abdel-Hafez [118] and Kazem and Hadi [119].

As noted by Verdial and others [120], In the event that the freesia plant was sprayed with licorice extract, this increased the height of the plant, in addition to the number of flowers.

According to the results of the analysis mentioned in **Figure 9** that the components of licorice root extracts contain amino acids and high concentrations, such as Thereonin, which leads to its great importance in resistance to cultivated plants, as well as the active substance Apigenen, which was detected with licorice extract, which may have the greatest effect in increasing Dry weight. Amino acids are used in the process of regulation and balance in the root zone when the plant is exposed to a deficit in water or high salts in the root zone and glycine acid is the one that works to maintain the osmotic balance between the cytoplasm and the gap Subbarao et al. [121].

CMV virus affects the amount of yield by reducing the percentage of flowering of plants, causing distortions to flowers and leaves and reducing the proportion of chlorophyll, which results in reducing the size of fruits.

The explanation for the superiority of *S. platensis*, *P. fluorescens* and licorice extract is due to the possibility of a joint effect by stimulating plant growth and resisting the negative effect of virus infection, and may be due to the compounds that each contains separately and their effects, and that the algae work to increase vegetative growth as a result of their containment. On growth regulators and stimulating resistance, which leads to a higher percentage of nutrient absorption and an increase in leaf area, which in turn is reflected in the increase in plant weight, and this is consistent with what Crouch et al. [122].

4.2.4 Effect of compounds extracted from medicinal plants on virus inhibition and plant growth parameters

The results (**Figures 8–10**) obtained from the analysis of amino acids and the active compounds of Albizia leaf extracts, licorice and sespan showed the presence of acids: Aspartic acid, Glutamic acid Serine, Histiden, Therionine, Argnine, Tyrosine, Cystine in Albizia plant, as well as: Aspartic acid, Glutamic acid Serine, Therionine,

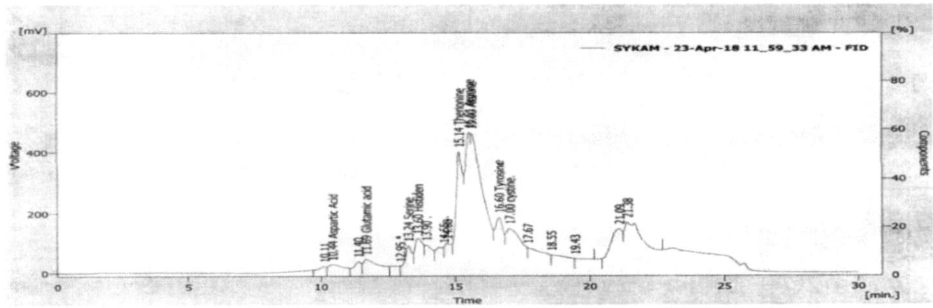

Result Table (Uncal - SYKAM - 23-Apr-18 11_59_33 AM - FID)

	Reten. Time [min]	Area [mV.s]	Height [mV]	Area [%]	Height [%]	W05 [min]	Compound Name
1	10.112	175.293	10.766	0.3	0.5	0.26	
2	10.440	492.039	15.671	1.0	0.7	0.58	Aspartic Acid
3	11.404	332.475	22.648	0.6	1.0	0.26	
4	11.688	808.825	29.886	1.6	1.3	0.46	Glutamic acid
5	12.948	48.142	2.852	0.1	0.1	0.10	*
6	13.244	1092.065	69.410	2.1	3.1	0.30	Serine .
7	13.596	1875.715	93.010	3.7	4.2	0.38	Histiden
8	13.904	1381.326	69.789	2.7	3.1	0.38	.
9	14.552	1198.291	59.795	2.3	2.7	0.35	..
10	14.684	1337.242	72.899	2.6	3.3	0.32	
11	15.140	6473.151	373.369	12.6	16.8	0.30	Therionine
12	15.512	6259.813	435.984	12.2	19.6	0.27	Argnine
13	15.600	12976.229	431.934	25.3	19.4	0.48	Alanine
14	16.604	3250.720	152.782	6.3	6.9	0.43	Tyrosine
15	17.004	4286.765	114.048	8.4	5.1	0.72	cystine.
16	17.672	1757.426	47.065	3.4	2.1	0.84	
17	18.552	666.430	19.658	1.3	0.9	0.72	
18	19.428	118.294	5.018	0.2	0.2	0.68	
19	21.088	2540.976	91.996	5.0	4.1	0.46	
20	21.384	4252.739	109.760	8.3	4.9	0.65	
	Total	51323.957	2228.341	100.0	100.0		

Figure 8.
Detection of the presence of amino acids and their values in Albizia leaves extract A. lebbeck.

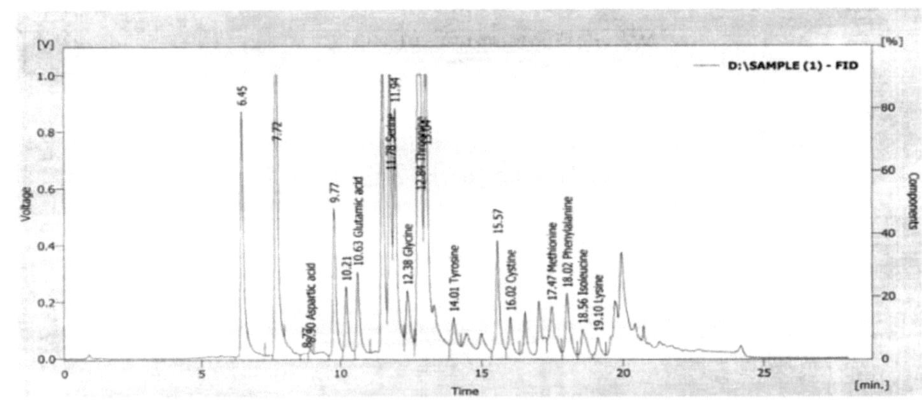

Result Table (ESTD - D:\SAMPLE (1) - FID)

	Reten. Time [min]	Response	Amount [uL]	Amount [%]	Peak Type	Compound Name
1	6.452	7350.263	0.000	0.0		
2	7.716	8954.984	0.000	0.0		
3	8.772	47.517	0.000	0.0		
4	8.900	113.937	4.701	0.9	Ordnr	Aspartic acid
5	9.772	4093.942	0.000	0.0		
6	10.208	1959.351	0.000	0.0		
7	10.632	2271.122	67.674	13.2	Ordnr	Glutamic acid
8	11.776	6112.431	62.529	12.2	Ordnr	Serine
9	11.936	5766.347	0.000	0.0		
10	12.384	1442.747	9.660	1.9	Ordnr	Glycine
11	12.836	11593.081	125.945	24.6	Ordnr	Threonine
12	13.040	11549.287	0.000	0.0		
13	14.012	861.197	10.504	2.0	Ordnr	Tyrosine
14	15.572	3248.922	0.000	0.0		
15	16.020	1024.908	17.012	3.3	Ordnr	Cystine
16	17.472	1406.384	75.005	14.6	Ordnr	Methionine
17	18.016	2042.658	66.342	12.9	Ordnr	Phenylalanine
18	18.560	1086.537	50.384	9.8	Ordnr	Isoleucine
19	19.104	587.899	22.743	4.4	Ordnr	Lysine
	Total		512.499	100.0		

Figure 9.
Detecting the presence of amino acids, and values at the root of licorice extract G. glabral.

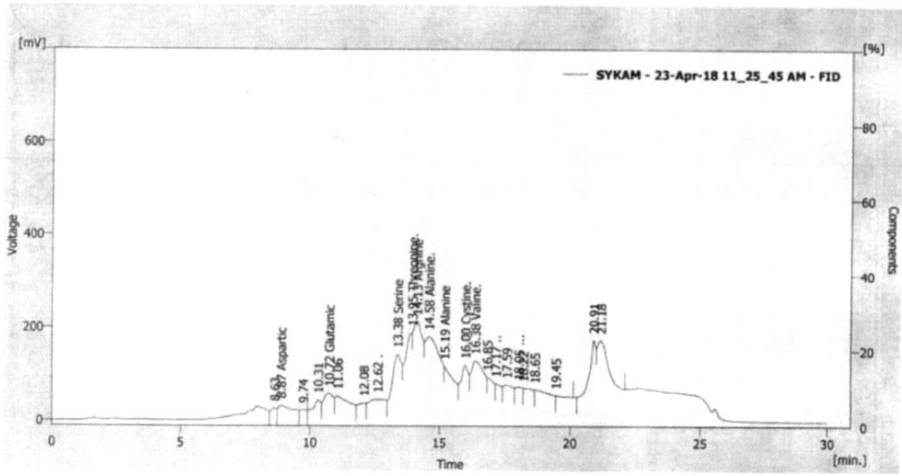

Result Table (Uncal - SYKAM - 23-Apr-18 11_25_45 AM - FID)

	Reten. Time [min]	Area [mV.s]	Height [mV]	Area [%]	Height [%]	W05 [min]	Compound Name
1	8.632	69.325	6.439	0.2	0.5	0.19	
2	8.872	248.771	11.887	0.7	0.9	0.32	Aspartic
3	9.736	7.211	0.607	0.0	0.0	0.22	
4	10.312	294.610	20.436	0.9	1.5	0.25	
5	10.724	821.633	33.588	2.4	2.4	0.45	Glutamic
6	11.060	815.095	26.999	2.4	2.0	0.54	
7	12.080	179.680	8.849	0.5	0.6	0.39	
8	12.616	653.887	15.521	1.9	1.1	0.74	.
9	13.384	2512.265	110.045	7.4	7.9	0.39	Serine
10	13.948	2779.395	154.526	8.2	11.2	0.38	Threonine.
11	14.132	4437.083	178.273	13.1	12.9	0.46	Argnine
12	14.584	5611.450	144.335	16.6	10.4	0.77	Alanine.
13	15.192	1896.059	79.110	5.6	5.7	0.55	Alanine
14	16.004	1601.167	79.686	4.7	5.8	0.39	Cystine.
15	16.384	2878.184	87.615	8.5	6.3	0.66	Valine.
16	16.848	861.307	51.197	2.5	3.7	0.32	
17	17.172	552.412	36.674	1.6	2.6	0.28	..
18	17.592	822.672	33.127	2.4	2.4	0.45	
19	18.052	515.141	27.698	1.5	2.0	0.32	...
20	18.224	567.760	23.863	1.7	1.7	0.43	
21	18.652	602.355	19.204	1.8	1.4	0.64	
22	19.452	86.089	5.136	0.3	0.4	0.46	
23	20.908	1742.040	116.360	5.1	8.4	0.22	
24	21.176	3305.922	113.340	9.8	8.2	0.46	
	Total	33861.514	1384.515	100.0	100.0		

Figure 10.
Detection of the presence of amino acids and their values in the extract of sespan leaves P. aculeate (L).

Tyrosine, Cystine, Methionine, Phenylalanine, Isoleucine, Lysine in the Licorice plant. In addition to: Aspartic, Glutamic, Serine, Therionine, Argnine, Alanine, Valine, and Cystine in the sespan. It has shown that the active compounds in Albizia leaf extract are Keamferol and Apigenen in licorice and Gallic acid, Qurcetine, Ellagic acid in Sespan.

The highest and best percentage of licorice, represented by the amino acid Therionine, was 24.6%, as it differed from the rest of the amino acids, as the amino acid Argnine for Spanish was 12.9%, the amino acid Argnine for Albesia 19.6%, and the active substance for licorice represented by Apigenen, which gave the highest concentration of 76.053 compared to Along with the rest of the extracts, the reason for this may be attributed to their effect on inhibition of the virus and increasing the growth parameters of leaf area, chlorophyll, plant height, dry vegetative and root system, as well as reducing the severity of infection in plants because amino acids are characterized by several physiological roles in plants and are of great importance to stimulate growth and maintain The pH of cells, and since amino acids contain two acid and basic groups as well, they lead to the expulsion of ammonia from cells and are also a store of carbon and energy.

Also, amino acids are distinguished by their ability to balance nutritional functions due to their effect on plant growth and yield due to the improvement of the original infrastructure in the cell, especially the plastids in the tissues, which improves the photocatalytic efficiency and leads to the production of more new cells that are reflected in the increase in plant height, The leaf area, as well as the yield and its components, as well as the amino acids have a role in stimulating the activity of a group of proteins responsible for enzymes for protein synthesis [123]. Free amino acids are an essential nitrogen source that is involved in building proteins and enzymes and providing energy that stimulates root and vegetable growth as well [124].

The abundance of amino acids has a great role in decreasing the osmotic voltage through which the water stress of the cell is reduced, and thus helps the cells in the process of drawing water and a set of nutrients from the medium and helps to increase the vegetative growth of plants [125]. 2002 that amino acids play a role as osmotic regulators as well as they regulate the process of ion transport, open stomata and expel toxins, a group of heavy metals and have a major role in the activity and building of enzymes.

5. Conclusions

1. The isolation of the virus that caused severe mosaic symptoms on cucumber plants with its different varieties in Al-Huwaish region is related to CMV virus, which genetically approximates the Pakistani isolate that was used in the molecular test for the virus.

2. The treatment of G. glabral root extract and *S. platensis* algae powder proved that it had a pathogenic effect on the CMV virus.

3. The ability of P. florescence to induce systemic resistance of different varieties of CMV-infected cucumber.

4. The triple complementary treatment (G glabral extract, *P. fluorescens* and *S. platensis*) was superior to the rest of the treatments in reducing the damage caused by CMV infection and raising some growth and yield parameters.

5. It has been proven by analyzing the amino acids of G.glabral licorice root extract that it contains the amino acid Therionine and the active substance Apigenen, which has the effect of inhibiting the CMV virus.

Author details

Maadh Abdulwahab Alfahad
Tikrit University, Iraq

*Address all correspondence to: maadhdft@gmail.com
and maadh.alfahd@tu.edu.iq

References

[1] Lower ، R.L. and M.D. Edwards (1986). Breading Vegtable crops .A.W. Publishing co. Inc. Westport ، Connecticut ،U.S.A.

[2] Ferreira S.A and Boley ، R.A. (1992) . CMV. University of Hawaii at Manoa.

[3] Zitter; T.A. (2004). Pepper Diseases Control-It Starts With the Seed. Cornell University،Vegetable Md Online، htt:// Vegetablemdonline.ppath.Cornell.edu.

[4] Agrios G.N (2005). Plant pathology 5th Ed Elsevier ، 922 p

[5] Hammerschmidt، R. ، (1999a). induced disease resistance : how do induced plants stop pathogen? Physiology molecular plant pathology 55: 77-84.

[6] Hammerschmidt، R. ، (1999b). phytoalexius: what have we learned after 60 years? Annual reiew phytopathology 37 : 285-306.

[7] Durrant، W.E. and X. Dong (2004). Systemic Acquired Resistance. PhytoPathology 42: 185-209.

[8] Walters; D.R.; D. Walsh; A. C. Newton and Lyon G.D.; (2005). induced resistance for plant disease control: Maximizying the efficacy of resistance elicitors. Phytopathology; 95: 1368-1373.

[9] Yusuf, Basma Dabab Aayed (2018). Molecular diagnostics and biological resistance of local isolation of ToMV tomato mosaic virus on cold pepper plants. Master Thesis - College of Agriculture - Tikrit University.

[10] Hull، R. (2002). Matthews Plant Virology. Fourth edition. Academic press، London ، Uk. pp. 1001.

[11] Shi، B.J.، P. Palukaitis and R.H. Symons. (2002). Differential virulence by strains of CMVis mediated by 2b gene. The American Phytopath. Society. 15(9): 947-955.

[12] Gibbs، A.J. and B.D Harrison(1970). Cucumber Mosaic Virus. C.M.I / A.A.B. Description of plant Viruses No. 1. Ferrylane. Kew ، Surrey ، England.

[13] Dolores، L. M.، and A. T. Makiling. (1997). Host range and symptomatology of chili veinal mottle virus (CVMV.) and CMV. isolated from pepper، *Capsicum anuum L.*

[14] Grescenzi، A.، A. Ragozzino، and A. Alioto (1994). Damage to eggplant by CMV(CMV.) Campania latium. Petria 4 (2): 141-149.

[15] Palukaitis P، and F. Garcia-Arenal. (2003). Cucumoviruses. Adv Virus Res 62: 241–323.

[16] Moury، B. (2004). Differential selection of genes of CMV subgroups. Mol. Biol. Evol. 21(8): 1602-1611.

[17] Roossinck ، M.J. (2001). CMV، a model for RNA virus evolution. Molecular plant pathology ، 2 (2) : 59-63

[18] Smith ، T.J. and K.L. Perry. (2000). The structure of CMVand comparison to cowpea chlorotic mottle virus. Journal of Virology ، Vol. 74 ، No. 16 ، P. 7578-7586.

[19] Soleimani، P.، Mosahebi، G.، Habibi، M.K.، (2011). Identification of some viruses causing mosaic on lettuce and characterization of Lettuce mosaic virus from Tehran Province in Iran. Afr. J. Agric. Res. 6 (13)، 3029–3035.

[20] Zitter، T. A. and Murphy، J. F. (2009). Cucumber mosaic virus. The Plant Health Instructor، 10، 1094.

[21] Raj، S.K.، A. Srivastava، G. Chandra and B.P. Singh.(2002). Characterization

of CMVisolating gladiolus cultivars and comparative evaluation of serological molecular methods for sensitive diagnosis. Current Science، 83 (9) : 1132-1137.

[22] Roossinck، M. J. (2002) Cucumoviruses (Bromoviridae)-general features. In: growth-promoting rhizobacteria systemically protect *Arabidopsis thaliana* against CMVby a salicylic acid and NPR1-independent and jasmonic acid-dependent signaling pathway. Plant Journal 39،381.

[23] Sill ، W.H.; Jr.; W.C. Burger ، M.A. Stahmann ، and J.C. Walker ، (1952). Electron Microscopy of Cucumber virus 1. Photopathology ، 42 : 320-422.

[24] Scott ، Howerd. (1963). Purification of CMV. Virology ، 20 : 103-106.

[25] Smith ،K.M.(1972). Plant Virus diseases Atext book. 3rd Edation. Longmans ، Group Ltd ، P : 234-252.

[26] Mathews ،R.E.F.(1981). Plant virology New York ، Academic Press. P : 247-250 ، 263 ،378 ، 405 609 ،619 ،660 ،238 ،797.

[27] Maramorosch، K. and Kerry F. Harris (1981). Plant dis. & Vector ،Ecology & Epidemiology. Academic press. :5-10 ، 17 ، 21-22 ، 40 ، 43،51-52 ،225 ،263 ،272-274.

[28] Shawkat, Abdullatif Bahjat (1982). Plant viruses, their properties. The diseases that cause them and their resistance. Dar Al Kutub Directorate Press for Printing and Publishing - University of Mosul. Pages: 112-113. 165-168 - 220-225.

[29] Price، W.C (1940). AM .J. Bot 27: 330. Cited by Frank. eteel. C .no. 213 in Description of plant Viruses. C.N.if، Assoo. APP1. Biologists. Kew Sueery England.

[30] Nelson. M.R. and Donald M.Tuttle (1969). The epid –Biology of CMVand arid Climate. Phytopathology ، 59 : 849 ، 855.

[31] Verma; N.; B.K. Mahinghara; R. Ram and A.A. Zaidi. (2006). Coat protein sequence shows that CMVisolate from geraniums (*Pelargonium* spp.) belongs to subgroup II. Journal of Biological Sciences; 31:47-54.

[32] Maramoro sch. K. (1977). Cucumovirus (CMV) Group. The Atlas of Insect and Plant Viruses. Academic Press ، New York ، Sanfrans1 Sco ، London P : 303-306.

[33] El-Sayed ، A. S.، Ali ، H.E. and Jarjees ، M.M. (1979). Some cucurbit viruses in Iraq. 3rd Egypt. Phytopathol. Congress ، P. 196-216.

[34] Cerkauskas، R. R. (2004). Cucumber mosaic virus AVRDC. The World Vegetable Center. Taiwan.

[35] Francki ، R.I.B. Mossop ، D.W. ; and Hatta F. (1979). Cucumber Mosaic Virus. No 213 in Description of plant Viruses. C.M.I/Assoc.App1. Biologists. Kew.Surrey. England.

[36] Ferrira، S.A. and R.A. Boley (1992). Cucumber Mosaic Virus. Crop knpwled master (Plant diseae pathogen.Retrieved from the. 1-5. Website

[37] Wang، Y.; V. Gaba ; J. Yang ; P. Palukaitis and Gal-On A.; (2002). Characterization of synergy between CMV And Potyviruses In Cucurbit Host. Phytopatho; 92 : 51- 58.

[38] Price ، W.C. (1934). Isolation and study of some yellow strains of cucumber mosaic virus. Phytopathology. 24 : 743-761.

[39] Paradies، F.، M. Finetti-Sialer، D. Gallitelli، and M. A. Yilmaz. (2000). Partial characterization of cucumber mosaic virus isolates from citrus and

grapevine. Journal Pl. Pathology. 82 (2): 133-145.f

[40] Al-Fahad, Maadh Abdel-Wahab. (2012). Immunological diagnosis of ZYMV potyvirus and evaluation of its interaction with three pumpkin cultivars under open & protected cultivation system in Salah al-Din. Tik. Univ. Jour. of Agr.Scien..

[41] Hu، J. S.، Li، H. P.، Barry، M. Wang، and R. Hordan (1995). Comparison of dot blot، ELISA، and RT-PCR assays for detection of two CMVisolates infecting banana in Hawaii. Plant Disease، 79 (9): 902-906.

[42] Ahangaran Gh. Mosahebi Mohammadi ، Mkoohi Habibi ، S. Khezri ، and N. Shahraeen. (2009). Use of Rapid Serological and Nucleic Acid based methods for detection the soybean mosaic virus. J. Agric.sci.Technol. Vol. 91-97.

[43] Iannelli، D. ، M. Barba، D. L. Apice، G. Pasquini، and C. Noviello (1996). Cytofluorimetric method for the detection of CMV. Phytopathology 86 (9): 959-965.

[44] Al-Ani, Sgt. Akef, Leila Jabbar, Sabr, Mustafa Ali, Torhab, and Wa'a Khudair Hassan. (2009). The response of some watermelon cultivars to CMVunder field conditions. Journal of Agricultural Sciences. 40 (6) 1–8.

[45] Al-Ani، R.A.، M. A. Adhab، A.A. Ali، and N. H. Diwa.(2011). Zucchini yellow mosaic virus: characterization and management in Iraq. International Journal Curr.، Res 11(3) : 220-224.

[46] Kumar، S. Udaya Shankar، AC.; Nayaka، SC، Lund OS، Prakash. HS. (2011). Detection of Tobacco mosaic virus and Tomato mosaic virus in pepper and tomato by multiplex RT-PCR. Letters in Applied Microbiology. 53:359-363.

[47] Akram W.، and T. Anjum. (2011). Use of bio agents and synthetic chemicals for induction of systemic resistance in tomato against diseases. Journal of Agricultural science and soil science 1(8): 286-292.

[48] Milus، E.A. and C.S. Rothrock. (1997). Efficacy of bacterial seed treatment for controlling pythium root of winter wheat. Plant Disease، 81:180-184.

[49] Hammerschmidt، R. and E. K. Dann. (1997). induced resistance to disease، P. 177-199. in: N.A. Rechcigl and Journal E. Rechcigl (eds.).

[50] Hemenway، C.، R.-X. Fang، W.K. Kaniewski، N.-H. Chua and N.E. Turner. (1988). Analysis. of the mechanism of protection in transgenic plants expressing the potato virus X coat protein or its antisense RNA، EMBO J. 7: 1273-1280.

[51] Al-Ani، R.A.، M. A. Adhab، A.A. Ali، and N. H. Diwa. (2011). Zucchini yellow mosaic virus: characterization and management in Iraq. International Journal Curr.، Res 11(3) : 220-224.

[52] Al-Fahad, Maadh Abdel-Wahab (2006). Diagnostic study, economic importance and resistance to Barly Yellow Dwarf Virus. PhD thesis. College of Agr. Univ. of Baghdad.

[53] Ahn،p; Park، K. and Hoekim، C. (2002). Rhizobactria-induced resistance perturbs viral disease progress and triggers defense related gene expression. Mol.cells، 13(2):302-308.

[54] Frindlender M، Inbar J & Chet I (1993) *Soil Biol Biochem* 25، 1211-1221.

[55] Kim، H. W. and Kim، B. K. (1999). Biomedical triterpenoids of *G. lucidum* (Curt.: Fr.) P. Karst. (Aphyllophoromycetideae). Int. J. Med. Mushrooms، ; 1: 121- 138.

[56] Bargabus ‹ R. L. Zidack ‹ N. K. Sherwood ‹ J. E. and Jacobsen ‹ B. J. (2002). Characterization of Systemic Rystemic Resistance in Sugar Beet Elicited By Anon Pathogenic. Phyllosphore Colonizing Bacillus My Coides. Biological Control Agent ‹ 66: 289-298.

[57] Al-Fahad, Maadh Abdel-Wahab. (1999). The use of some biological and natural factors in protecting tomato crop from infection by the *Bemisia tabaci* (Genn) tobacco fly and tomato leaf curl and yellowing virus (TYLCV). MSC Thesis, Basra Univ.

[58] Kurowski ‹ C. ‹ P. M. Remeeus (2006). ISIH – Veg comparative test : The Research Report 1- 2006- Psp. Noyn ‹ Switzeland: International Seed Federation.

[59] Kazempour‹ M.N. ‹(2004). Biological control of *Rhizoctonia solani* the caused agent of rice sheath blight by antagonistic bacteria in greenhouse and field condition. Plant pathology Journal 3 (2) :88-96.

[60] Ali‹ SH. K.; Saleh‹ A. M. (2012). Spirulina-An over view. International Journal of Pharmacy and Pharmaceutical sciences. 3(4): 9-15.

[61] Kewi‹ C.; David‹ L.; Keith‹ K.; William D. and Brett‹ A. N. (2011). Molecular classification of commercial Spirulina strains and identification of their sulfolipid biosynthesis genes. J. Microbiol biotechnol. 21(4): 359-365.

[62] Benson‹ J. M. (2008). Cyanobacterial toxins. Drugs. Sited by Kumar‹ V.; Tirumalai‹ P. S.; Abha‹ S.; Bhatnagar A. K. and Shrivastava J.N. (2013). Indo Global Journal of Pharmaceutical Sciences‹; 3(3): 212-223.

[63] Newman‹ D. J.; Cragg‹ G. M. and Snader‹ K. M. (2000). The influence of natural products upon drug discovery. Nat. Prod. Rep.‹ 17: 215-234.

[64] Kim H.G. (2006). In Ecology of Harmful Algae‹ E. Graneli‹ J. T. Turner (eds). Ecological Studies‹ Vol. 189‹ pp. 327-338.

[65] Tadros‹ M. G. and Macelroy‹ R. D. (1988). Characterization of Spirulina Biomass Y for CELSS Diet Potential. Alabama AEM University.P33.

[66] Juarez-Oropeza‹ M.A.; Mascher‹ D.; Torres ‹ P.V.; Farias‹ J.M.;Paredes-Carbajal‹ M.C. (2009). Effects of Spirulina on vascular reactivity. J. Med. Food‹ 12‹ 15–20.

[67] Capelli‹ B.; Cysewski‹ G.R. (2010). Potential Health Benefits of Spirulina: A Review of the Existing Literature; Cyanotech Corporation: Kailua-Kona‹ HI‹ USA‹ 2010.

[68] Blinkova‹ L. P.‹ O. B. Gorobets‹ and A. P. Baturo. (2001). Biological activity of Spirulina. Zh. Mikrobiol. Epidemiol. Immunobiol. 2: 114-118.

[69] Abd El-Motty‹ E. Z.; Shahin‹ M. M.; EI-shiekh‹ M. H. and AbdMigeed‹ M. M. (2010). Effect of algae extract and yeast application on growth ‹Nutritional status ‹ yield and fruit quality of Keitte mango trees. Agric. Biol. J. N. Amer. 1 (3): 421-429.

[70] Ray‹ B. (2004). In vitro anti-herpetic activity of sulfated polysaccharide fractions from Caulerpa racemosa. Phytochemistry. 65. 3151-3157

[71] Kim‹ S. k. and Karadeniz‹ F. (2011). Anti-HIV Activity of extracts and compounds from marine algae. In Advanced Food and Nutrition Research‹ vol. 64:.213-224.

[72] Bulter‹ T. and Hunter‹ A. (2006). Impact of seaweed extract on Turf grass growth and nutrition on a golf green to USGA specification. Acta Horticulturae (ISHS). 762: 81-90.

[73] Mishima، T.; Murata، M.; Toyoshima، H.; Fuji، M.; Nakajima، T.; Hayashi، T. and Kato، I. (1998). Inhibition of tumor invasion and metastasis by calcium spirulina (Ca-Sp)، a novel sulfated polysaccharide derived from blue-green algae□ *Spirulina Platensis*. Clin. Exp. Metastasis، 16(6): 541-50.

[74] Spenille، F.; Fiori، G.; Noferini، M. Sprocatti، M. and Costa، G. (2009). Perspective on the use of a seaweed extracts to moderate the negative effects of alternate bearing in Apple trees. Journal of special issue :131 137.

[75] Basma Dhabab Ayed* and Maadh Abd Al- Wahab Al-Fahad (2018). Molecular Identification and Biological Resistance to Local Isolate of the Tomato Mosaic Virus (ToMV) on the Sweet Pepper Plants. Journal Tikrit Univ. For Agri. Sci. Vol. (18) No.(3).

[76] WWW.iimsam.erg

[77] Gupta، ML. Janardhanan KK، Chattopadhyay A & Hussain A. (1990). Association of Glomus with palmarosa and its influence on growth and biomass production. Mycol. Res. 94: 561-563.

[78] Apablaza، G. E.، and C.C. Bernier (1972). Inhibition of tobacco mosaic virus infection by plant extracts. Can. J. Bot. 50:1473-1478.

[79] Ndung'u ، J.، (2008). Data collection and evaluation of Sesbania in southern Africa. http:/indrinfo.idrc.ca/Archive/corpdocs/114796/Data.htm.

[80] Kirtikar, K.R. and Basu, B.D. (1993). In: Blatter, E., Caius, J.F. and Mhaskar, K.S., Editors, Indian Medicinal Plants I-IV, Lalit Mohan Basu, Allahabad, India

[81] Raju N Patil, Ravindra Y Patil, Dheeraj Ahirwar (2010) STUDY OF SOME MEDICINAL PLANTS FOR ANTIDIABETIC ACTIVITY IN ALLOXAN INDUCED DIABETES. Pharmacologyonline 1: 53-60

[82] Davis، N.D.; Dinner، U.L. and Eldridge، D.W.(1967). Production of aflatoxin B and G by Aspergillus flavus in semi synthetic medium.Appl. Microbiol. 14:378-380.

[83] Mazzola، M.; Cook، R.J.; Thomashow،L.S.;Weller، D.M. and Pierson، L.S.(1992). Contribution of phenazine antibiotic biosynthesis to ecological competence of fluorescent Pseudomonas in soil Habitates.Appl.En. Mic.P.2616-2624.

[84] Bhat، G. S. and Chauhan، P. S. (2002). Provenance variation in seed and seedling traits of Albizia lebbeckBenth،" Journal of Tree Science. 21: 52–57.

[85] Faisal، M.، Singh، P.P. and Irchhaiya، R. (2012). Review on Albizia lebbecka potent herbal drug،" International Journal of Pharmaceutics. 3(5): 63–68.

[86] Kumar، A.; A.K.، Saluja; U.D.،Shah and A.V. Mayavanshi.(2007). Pharmacologica. potential of Albizzia lebbeck: A Review. Pharmacognosy Reviews, 1: 171-174.

[87] Nazneen Bobby، M.D.; Wesely، E.G. and Johnson، M. (2012). High performance thin layer chromatography profile studies on the alkaloids of Albizzia lebbeck. Asian Pacific Journal of Tropical Biomedicine. 51- 56.

[88] Abbas، H.H.; V. Mahnaz; K. Hamid; R. Sadeg; H.R. Mohsen; V. Motahareh،K. and Mohammed، R.A. (2006). Chemical composition and toxicity of *Melia azedarach L*

[89] Tyler، V. (1993).TheHonest Herbal، Pharmaceutical products press، Ny. pp. 197990.

[90] Abu Zaid, Shahat Nasr. (1986) Plants and medicinal herbs, National

Research Center, Cairo, Crescent Library House, Beirut, First Edition.

[91] Al-Ajili, Thamer Abdullah Zahwan. (2005) The effect of gibberellin GA3 and some nutrients on the production of glycyrrhizin and some other ingredients in the licorice plant Glycyrrhiza glabra, PhD thesis, College of Agriculture, University of Baghdad, Iraq.

[92] Ibrahim, "Diaa Khalil. (1985) Licorice plant, its components, its benefits. Journal of Science 29. 2 (3): 55-56, Iraq.

[93] Al-Janabi, Abd al-Basit Abbas Ali. (1984) Effects of various plant extracts on tobacco mosaic virus TMV. Master Thesis, College of Agriculture-University of Baghdad.

[94] Fukui H، Gotok (1988).Two antimicrobial flavanones from the leaves of *Glycyrrhiza glabra*.Chem. Pharm. Bull.; 36(10):4174–4176.

[95] Kovalenko. P.G. and SS. Maliuta (2002). Secondary metabolite production from cell culture of *Glycyrrhiza glabra* and potentilla alba for production of radio protective compounds، NAS of NA، zabolotn. Go 150، Kyiv – 143، Ukraine

[96] WHO. (1999). World Health Organization. www.who.int

[97] Bhalla, S., Gupta, K. Lal, B., Kapur, M.L. and Khetrapal, R.K. 2008. Efficacy of various non-chemical methods against pulse beetle, Callosobruchus maculatus (F.). ENDURE International conference, Diversifying Crop Protection, October, 12-15, France.

[98] Qasim, Nabil Aziz, Ali, Hamid Mahmoud, (2012). Practical Plant Viruses, Mosul University Press, page 466.

[99] Scriver CR، Beaudet AL، Valle D، Sly WS، Childs B، Kinzler KW، Vogelstein B، eds. The Metabolic and Molecular Bases of Inherited Disease. 8th ed. New York، NY: McGraw-Hill، Inc; 2001;1665-2105

[100] Badifu، G.I.O.، (1991). Chemical and physical analysis of oils from four species of cucurbitaceae ، J. Am. Oil Chem. Soc. 68 : 428-432.

[101] AOAC (1995). Official methods of analysis (16th Ed.). Washington، DC : Association of Official Analytical Chemists

[102] Mckinney، H.H.(1923). Influencee of soil temperature and moisture on infection of wheat seedling by Helmiathosporium sativum. J. Agric. Res.،26:195-217.. 130 136.

[103] Shomaila Igbal ، Hussain Shah ،Muhammad Ashfaq and Ajmal Khan Kassi ، (2017) ،Standardization of PCR protocol to check diversity among Pakistani isolates of (CMV) infecting chilli pepper، Journal of Ento. and Zoology Studies ; 5(3): 729 _735.

[104] Jeannens, A. and Teta, M. 1971. chromotographic estimation of glycyrrhizic acid concentration in a licorice especiality planta med phytother.5:214-223.

[105] Maurhofer, M., Hase, C., Meuwly, P., Metraux, J-P., and Defago, G. 1994. Induction of systemic resistance to tobacco necrosis virus by the root-colonizing *Pseudomons flurescens* strain CHA0: Influence of the gacA gene and pyoverdine production. Phytopathol. 84:139-146.

[106] Abbassy، M. A.; Marei، G. I.; Selwan، M. H. (2014). Antimicrobial Activity of Some Plant and Algal Extracts. *International Journal of plant* & soil science. 3 (10): 1366-1373.

[107] Usharani، G.; Srinivasan، G.; Sivasakthi، S.; Saranraj ، P. (2015). Antimicrobial Activity of Spirulina

platensis Solvent Extracts Againts pathogenic Bacteria and Fungi. Advances in Biological Research. 9(5): 292-298.

[108] Murphy‹ J. F.‹ Reddy‹ M.S.‹ Ryu‹ C.M.‹ and Kieopper R. Li. (2003). Rhizobacteria mediated growth promotion of tomato leads to protection against Cucmber mosaic virus. Phytopathology. 93:1301-1307. NASA and Spirulina (Argentina):

[109] Ryu‹ C. M.‹ Farag‹ M. A.‹ Hu‹ C. H. ‹ Reddy‹ M. S.‹ Kloepper‹ J. W.‹ and Paré‹ P. W.‹(2004)‹ Bacterial volatiles ISR in Arabidopsis. Plant Physiol. 134: 1017-1026

[110] El-Dougdoug‹ K. A.; Ghaly‹ M.F. and Taha‹M.A. (2012). Biological control of CMVby certain local Streptomyces isolates:Inhibitory effects of selected five Egyptian isolates. International Journal of Virology. 8(2): 151-164

[111] Figueiredo MVB‹ Seldin L‹ Araujo FF‹ Mariano RLR (2010) Plant growth promoting rhizobacteria: fundamentals and applications. In: Maheshwari DK (ed) Plant growth and health promoting bacteria. Microbiology monographs 18. Springer‹ Berlin‹ 21–43

[112] Mao‹ T. K; Water‹ V. de. And Gershwin‹ M. E. (2005). Effect of Spirulina-based dietary supplement on cytokinen production from allergic rhinitis patients. Journal Medical Food. 8(1): 27-30.

[113] Hillel, D. (2005). Plant Growth Promoting Bacteria. Elsevier, Oxford, U. K., 103-115. http://Spirulina-astronautfood.blogspot.com/2009/nasa-andspirulina.Html.http://Spirulina-a stronautfood.blogspot.com/2009/nasa-a nd spirulina. Html.

[114] Hassan, Ahmed M. and Jumaa, Farouk F. (2013). Effect of spraying with some growth stimulants on the vegetative growth characteristics of seedlings. 5(4)244-225.

[115] Al-Ani, Abdul-Ilah Mukhlif. (1980). The effect of spraying licorice extract on the growth and yield of tomato. Rafidain Agriculture Journal. Iraq.

[116] Bidwell‹ R.G.S.(1979). Plant Physiology. 2nd ed. Macmillan Pub. Co. Inc. ‹N. Y.

[117] Ashmead, H.D. 1986. The absorption mechanism of amino acid chelates by plant cells. In: Foliar feeding of plants with amino acid chelates (Ashmead, H.D.; H.H. Ashmead; G.W. Miller and H.H. Hsu, Noyes Publications, Park Ridge, New Jersey, USA), 219-235.

[118] Abdel Hafez, Ahmed Abu Al-Yazid. (2011). The use of algae and seaweed to improve the growth of horticultural crops. Shams Agricultural Magazine. Issue (122). Number (5). faculty of Agriculture. Ain Shams University. Egyptian.

[119] Kazem, Ahmad Adnan, Hadi, Akram Abd al-Kazim (2015). Effect of spraying with seaweed extract and humic acid on growth indicators of persimmon seedlings, Diospyrus Kaki L. Al-Furat Journal of Agricultural Sciences. 5 (4): 272-279.

[120] Verdial‹ M.F; T. Neto; K. Minami; J. A. S. Filho; P. J. Christoffoleti; F. V. Scarpare;J. F. Barela;J. S. D Aguial and R. A‹ Kluge (2007). Vernalization on five cultivars of strawberry. Ciencia Rural‹ Santa Maria ‹37(4):976-981.

[121] Subbarao‹ G.V.‹ C. Johansen‹ A.E. Slinkard‹ R.C.N. Rao‹ N.P.Saxena‹ and Y.S. Chauhan‹ 1995. Strategies for improving drought resistance in grain legumes. Crit Rev Plant Sci.‹ 14: 469-523.

[122] Crouch, I.J., Beckett, R.P., Van Staden, J., 1990. Effect of seaweed concentrate on the growth and mineral nutrition of nutrient stress lettuce. Journal of Applied Phycology 2, 269–272.

[123] Mladenova، Y.I.، P. Maini.، C. Mallegni.، V. Goltsev.، R. Vladova، K. Vinarova and S. Rotcheva. (1998). Siapton، an amino acidbased biostimulant reducing osmostress metabolic changes in maize. Agro-Food Ind. Hi-Tech.، 9: 18-22.

[124] Abdel-Aziz and L. K. Balbaa. (2007). Influence of tyrosine and zinc on growth، flowering and chemical constituents of Salvia farinacea plants. J. of Applied Sci. Res.،3(11): 1479 – 1489.

[125] Abu Dahi, Youssef M., Muayyad Ahmad Al-Younes (1988). Plant Nutrition Guide. Ministry of Higher Education and Scientific Research. Baghdad University. Directorate of Dar al-Kutub for printing and publishing/ University of Mosul.

Chapter 12

Current and Emerging Pests and Diseases of Cucumber (*Cucumis sativus* L.) in Africa

Olawale Arogundade, Titilayo Ajose,
Joshua Olumide Matthew and Itunu Adeyemi Osijo

Abstract

The place of cucumber (*Cucumis sativus* L.) in Africa was considered insignificant for years due to its previously assumed limited uses. However, it has now gained recognition as one of the important market vegetables in the tropics because of the continued awareness of the numerous health benefits attached to its consumption. This has progressively affected its cultivation and thereby, increased outbreak of diseases and insect pests of the crop. High incidence of insect and disease infestations occur in *Cucumis sativus* L. due to extreme temperatures, heavy rainfall and high humid condition resulting there from, causing huge losses through reduction in yield, lowered quality of harvested produce and increased cost of production. More than 30 pests and diseases are known to contribute to the losses recorded on the crop in Africa, some of which will be discussed in this chapter.

Keywords: infestation, insect, losses, outbreak, tropics, yield

1. Introduction

Cucumber (*Cucumis sativus* L.) is considered the fourth most essential vegetable worldwide, perhaps due to its great nutritional, medicinal and economic potential [1, 2]. In terms of world total production, cucumber is rated an important cucurbit alongside watermelon (*Citrullus lanatus* L.) and melon (*Cucumis melon* L.) [3]. Low yield and insufficient use of the product previously contributed to ranking the crop insignificant in Africa [4]. At present, the demand for cucumber is on the increase daily in Africa because of the continued campaigns about the numerous benefits of the crop [5].

Despite the increasing relevance of cucumber, production is seriously constrained by many factors which include scarcity of suitable planting materials, limited access to capital, climatic conditions, plant pests and diseases, among others [5]. More than 40 diseases caused by viral, bacterial, fungal and nematode pathogens severely affect the cultivation and production of *C. sativus* L. [6]. Insect pests are reported to consume crops sufficient to feed an additional one billion people on a worldwide basis, thereby placing much importance on identifying and managing them prior to infestation [7].

Like other cucurbits, the most common pests of *C. sativus* L. are spotted cucumber beetle (*Diabrotica undecimpunctata*) and striped cucumber beetle

(*Acalymma vitatum*) [8]. Other pests identified on cucumber fields in Africa include Flea beetle (*Phyllotreta cruciferae*), Hadda beetle (*Epilachna vigintioctopunctata*), banded cucumber beetle (*Diabrotica balteata*), squash bug (*Anasa tristis*) and squash vine borer (*Melittia cucurbitae*). Additionally, *C. sativus* L. harbours a number of aphids which include melon aphid (*Aphis gossypii*), cowpea aphid (*Aphis craccivora*), potato aphid (*Macrosiphum euphorbiae*) and green peach aphid (*Myzus persicae*), which serve as common vectors of important viral diseases of *C. sativus* L. [9].

In Africa, cucumber is largely affected by viruses belonging to three genera; namely *Potyvirus*, *Cucumovirus* and *Crinivirus* [10]. *Zucchini yellow mosaic virus* (ZYMV), *Watermelon mosaic virus* (WMV), *Moroccan watermelon mosaic virus* (MWMV), *Papaya ringspot virus* (PRSV), *Cucumber mosaic virus* (CMV) and *Beet pseudo-yellows virus are some of the viruses* that cause severe symptoms to cucumber [11]. Other viruses of minor importance include *cucurbit aphid-borne yellows virus*, *cucurbit yellow stunting disorder virus*, *melon necrotic spot virus*, *squash mosaic virus* and *tomato spotted wilt virus* [12]. Plant virus infections result to reduction in plant growth, lower yields, compromised fruit quality, reproductive difficulty, increased susceptibility of the host plant to other pathogens and economic losses to farmers [13].

Several fungi attack seedlings of cucumber, causing appreciable losses, especially under favourable environmental conditions [14]. In Africa, many pathogenic fungi such as *Alternaria tenuis*, *Alternaria alternata*, *Fusarium equisett*, *Fusarium solani*, *Aspergillus* spp., *Phytophthora capsici*, *Penicillium oxalicum*, *Bipolaris* spp., *Botrytis cinerea*, *Cladosporium tenuissimum*, among others, have been associated with rotting of cucumber fruits [15], causing reasonably high post-harvest losses [16, 17]. Downy mildew, powdery mildew and anthracnose also cause substantial losses. Scab affects aboveground plant parts, including the fruits.

Root-knot nematodes are prevalent destructive pathogens of *C. sativus* L. because even at low levels, high yield losses result [18]. Other nematodes include species of *Rotylenchus*, *Benololaimus*, *Pratylenchus*, *Paratylenchus* and *Trichodorus*. Angular leaf spot is the most common bacterial disease of *C. sativus* L., although it affects all cucurbits. Bacterial wilt disease has also been reported in Africa [19].

The increasing trend of local and international movement of seedlings, cuttings and fruits enhance the risk of introducing new pathogens and vectors, where applicable into new areas. Changing climate conditions can contribute to a more successful spread of pathogens and establishment of such organisms in areas previously unfavourable to their existence. A holistic disease management approach which requires the use of cultural, mechanical, biological and chemical methods is needful to mitigate the negative impact of diseases, insect and weed pests on productivity, hence, the importance of this review on the current and emerging pests and diseases of *C. sativus* L. in Africa.

2. Viruses of cucumber

2.1 Genus: *Potyvirus*

The genus *Potyvirus* is one of the largest genera of plant viruses infecting several economically important cucurbits worldwide [20]. Potyviruses are non-enveloped flexuous filamentous viruses of about 680–900 nm long and 11–15 nm wide, harbouring a monopartite genome consisting of a single positive-sense RNA covalently linked to a viral protein genome at one end and a polyadenyl tail at the other end [11, 21]. Typical symptoms induced by *potyviruses* include mosaics on leaves, leaf distortion, leaf reduction, fruit deformation, yellowing and wilting [22]. The major *potyviruses* that affect *C. sativus* L. in Africa include *Zucchini yellow mosaic virus* (ZYMV), *Watermelon mosaic virus* (WMV), *Moroccan watermelon mosaic virus* (MWMV and *Papaya ringspot virus* (PRSV).

2.1.1 Zucchini yellow mosaic virus (ZYMV)

Zucchini yellow mosaic virus (ZYMV) naturally infects cucurbits, mostly cultivated species, a few ornamentals and weeds. There are reports of ZYMV in South Africa, Egypt, Nigeria, Sudan, Tunisia, Mali, Madagascar and Morocco [23, 24]. ZYMV is spread by aphids in a non-persistent manner [22]. Among 26 aphid species capable of transmitting ZYMV, *Aphis gossypii* Glover, *Aphis craccivora* Linnaeus, *Macrosiphum euphorbiae* Thomas and *Myzus persicae* Sulz have been identified as more efficient vectors. Mechanical and seed transmission have also been reported [25]. Symptoms produced on infected plants include vein clearing, yellow mosaic, leaf deformation with blisters, misshaped fruits and stunted growth (**Table 1**).

2.1.1.1 Management of ZYMV

Planting of available resistant/tolerant varieties to ZYMV is advisable. Growing taller non-susceptible companion crop, cross- protection to control severe ZYMV isolate and spraying of insecticides are also good measures against the virus.

2.1.2 Watermelon mosaic virus (WMV)

Watermelon mosaic virus (WMV), formerly WMV-2 has a wide host range which include *Cucurbitaceae*, *Apiaceae*, *Chenopodiaceae*, *Fabaceae*, *Malvaceae*, *Orchidaceae* and several weeds [33, 34]. WMV has been reported in such countries as Nigeria, South Africa and Tanzania in Africa [35, 36]. About 35 species of aphids transmit WMV in a non-persistent manner although *A. gossypii*, *A. craccivora* and *Myzus persicae* are regarded as more efficient vectors. Recent studies have also established the possibility of WMV transmission through seed [37]. Symptoms of WMV is dependent on the host and the isolate and these include vein-banding, mosaic, leaf deformation, fruit discolouration and distortion [38].

2.1.2.1 Management of WMV

Eradication of weeds and alternative host helps to reduce the incidence of WMV. Close monitoring of field for timely intervention in the event of WMV disease outbreak is helpful. The use of plastic mulches also reduces insect infestation which, otherwise would transmit the virus.

2.1.3 Moroccan watermelon mosaic virus (MWMV)

Moroccan watermelon mosaic virus was first reported in Morocco in the year 1972 as a strain of WMV, causing severe diseases in various cucurbits [39]. Subsequently, MWMV was discovered as a distinct potyvirus species distantly related to Papaya ringspot virus (PRSV) based on biological and serological properties [40]. The host range of MWMV is limited to members of *Cucurbitaceae* family and papaya. The geographic distribution of the virus in Africa spread through Niger, Cameroon, Nigeria, South Africa, Tunisia, Tanzania, Congo and Zimbabwe [41–43]. *A. gossypii* and *Myzus persicae* transmit the virus in a non-persistent manner. Symptoms associated with MWMV infection include mosaic, severe leaf and fruit deformation, wilting and dark-green blistering [41].

Causal	Disease	Picture	Reference	Causal	Disease	Picture	Reference
Virus	ZYMV	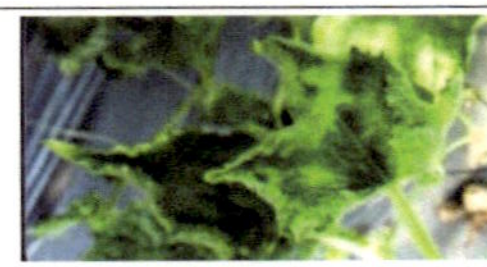	[26]	Bacteria	Angular leaf spot		[27]
Virus	WMV		[26]	Bacteria	Root Knot		[26]
Virus	MWMV		[28]	Nematode	Downy mildew		[29]
Virus	PRSV		[30]	Fungi	Powdery mildew		[26]
Virus	CMV	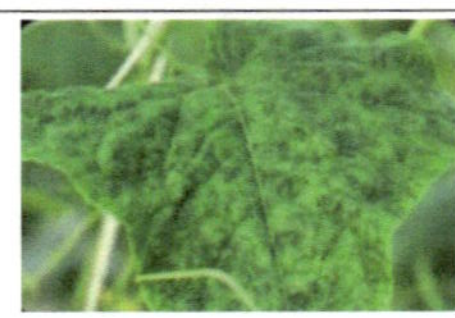	[26]	Fungi	Alternaria leaf blight		[26]

Causal	Disease	Picture	Reference	Causal	Disease	Picture	Reference
Virus	BPYV		[31]	Fungi	Fusarium wilt		[26]
Bacteria	Bacterial wilt		[26]	Fungi	Damping-off		[32]

Abbreviations: *ZYMV: Zucchini yellow mosaic virus; WMV: Watermelon mosaic virus; MWMV: Moroccan watermelon mosaic virus; PRSV: Papaya ringspot virus; CMV: Cucumber mosaic virus; BPYV: Beet pseudo-yellows virus.*

Table 1.
Symptom expression of some cucumber diseases.

2.1.3.1 Management of MWMV

Use resistant varieties for planting and avoid planting close to old cucurbit fields. Phyto-sanitation, use of virus-free planting materials and pesticide for control of the vectors have also been employed in the management of MWMV.

2.1.4 Papaya ringspot virus (PRSV)

Papaya ringspot virus (PRSV) has great economic importance for cucurbit and papaya cultivation worldwide. The designated biotype for cucurbits is PRSV-W. The virus has been reported on cucurbits in Tunisia [44], Egypt [45], Morocco [46] and South Africa [47]. Early season infection may lead to poor fruit set while late season infection may result to blotchy fruit. The virus can be transmitted through movement of farm workers and machinery from one place to another and by aphids. At initial stage, symptoms can appear as vein clearing of leaves followed by the development of dark-green mosaics. In cucumber, leaves are distorted along the margins.

2.1.4.1 Management of PRSV

Management of infection on cucurbits can be achieved by using resistant varieties for planting. Insecticide applications may also reduce aphid numbers in the field. Other practices may include the use of mulches and elimination of volunteer weed hosts.

2.2 Genus: *Cucumovirus*

Viruses in this genus have wide host range, infecting over 1200 plant species worldwide [48]. Virions are icosahedral particles of approximately 29 nm in diameter, made of subunits of single capsid protein numbered 180 [49, 50]. The genus consists of three linear positive-sense single stranded RNA molecules. The major *Cucumovirus* infecting *C. sativus* L. in Africa is *Cucumber mosaic virus* (CMV).

2.2.1 Cucumber mosaic virus (CMV)

The first report of cucurbits-infecting virus was obtained on *Cucumber mosaic virus* in 1916. CMV is of great importance in temperate and tropical regions of the world [51]. Report of CMV infection on cucurbits in Africa were established in Tanzania and South Africa [36]. Over 80 species of aphids in more than 30 genera transmit CMV in a non-persistent manner but *Aphis gossypii* and *Myzus persicae* are the most efficient [52, 53]. Rapid spread of the virus has been attributed to the attraction of CMV-infected plants to aphids [54]. Transmission through seed, parasitic weeds such as dodder *Cuscuta* spp. and mechanically have also been established [50, 55, 56]. CMV causes typical mosaic symptoms in cucumber which include mosaic on leaves or fruit, stunted growth, deformed fruit, yellow spot and wilting (**Table 1**).

2.2.1.1 Management of CMV

The use of certified seed for planting, regular weeding and disinfection of hands and farm tools are effective ways to manage CMV. Aphid population should also be kept under control through regular spraying with insecticide.

2.3 Genus: *Crinivirus*

The genus *Crinivirus* comprises of members with cross-banded flexuous particle, five gene nodule and infection that is restricted to the vascular tissues [57]. A typical member of this genus infecting cucurbits in Africa is *Beet pseudo-yellows virus*.

2.3.1 Beet pseudo-yellows virus (BPYV)

Beet pseudo-yellows virus has a broad host range including cucurbits, ornamentals and weeds. BPYV particles are approximately 12 nm wide and 1500–1800 nm long [58]. BPVY has been reported in South Africa [11]. The genome is made up of two linear positive sense, single stranded RNA of about 7.6 to 8 kb, both required for infectivity [57]. Typical BPVY symptoms may be confused with physiological disorders. Symptoms appear first on older leaves as yellow spots which develop into yellow blotchy raised areas between veins, which remain green. Subsequently, younger leaves become affected but the fruit remain intact.

2.3.1.1 Management of BPYV

Prevention of incidence of whitefly infestation, practising crop rotation and regular weeding are recommended ways of managing BPVY. Elimination of inter-cropping of old and young plants and good sanitation are also effective.

3. Bacterial diseases of cucumber

3.1 Genus: *Erwinia*

The genus *Erwinia* comprises rod-shaped bacteria that are plant-pathogenic and plant-associated [59]. Bacteria in this group are related to *Escherichia coli*, *Shigella*, *Salmonella* and *Yersinia*. A typical member of this genus which affects *C. sativus* L. is *E. tracheiphila* which causes bacterial wilt of cucurbits in general.

3.1.1 Bacterial wilt of Cucurbits

Bacterial wilt, caused by *Erwinia tracheiphila* and vectored by striped cucumber beetle (*Acalymma vittatum* F.), is one of the serious diseases threatening natural and wild cucurbit crops. Cucumber is one of the most susceptible to the disease [60]. Bacterial wilt remains a major disease of cucurbits in South Africa [61]. Transmission occurs when an infected beetle creates wound during the course of feeding on a plant and defecates therein. The bacterium moves by free water into the xylem, which spreads and multiplies it. Disease symptom appears first on younger plants and these include discolouration of stem tissues and wilting of some or all the parts of the affected plant [62].

3.1.1.1 Management of Bacterial wilt disease of cucurbits

Disease management relies on controlling cucumber beetles, mainly through insecticide applications. Practicing crop rotation and planting cultivars that are less attractive or susceptible to striped cucumber beetles are also helpful.

3.2 Genus: *Pseudomonas*

Members of the genus *Pseudomonas* are pathogenic aerobic bacteria that are naturally widespread in the environment [63]. Their wide occurrence has been attributed to their great deal of metabolic diversity and as a result, ability to colonise a wide range of niches. One of the most studied members of the group is *Pseudomonas syringae*. Angular leaf spot caused by *Pseudomonas syringae* pv. *lachrymans*, is common to almost every cucumber growing area of the world [64].

3.2.1 Angular leaf spot

Angular leaf spot is one of the serious diseases which primarily affects cucumber. The infection can affect the quality of fruit produced by inflicting up to 37 and 40 per cent reduction in fruit number and fruit weight, respectively in addition to rendering some fruits totally unmarketable [65]. At the initial stage, the symptoms appear on leaves in form of small, water-soaked lesions which later enlarge. Older lesions become angular as they enlarge and encounter veins. The infected tissues often dry and fall. The leaves are left with torn-irregular-shaped holes. In Africa, angular leaf spot disease was identified on cucumber seedlings raised from infected seeds in Egypt, with disease incidence of 98 per cent [66]. Different isolates of *P. syringae* pv. *lachrymans* from Egypt are also reported to induce wilting, besides the typical angular leaf spots, within 3 to 6 days of inoculation [67].

3.2.1.1 Management of Angular leaf spot

Angular leaf spot disease can be effectively managed by planting disease-free seeds. Additionally, planting field should have good drainage system to avoid any form of contact from neighbouring cucurbitaceous fields. Crop rotation with non-cucurbits and complete removal of crop remains after harvest are helpful.

4. Nematode diseases of cucumber

4.1 Genus: *Meloidogyne*

Members of the genus *Meloidogyne* have gained worldwide recognition as one of the major plant parasites constraining crops of primary economic importance [68]. *Meloidogyne* spp. are considered among the top five major plant pathogens and the first among the ten most important genera of plant parasitic nematodes in the world [18]. *M. arenaria*, *M. incognita* and *M. javanica* are reportedly dominant in Africa [69]. *Meloidogyne* spp. were among the nematodes found on cucumber field in Libya [70]. Aminu-Taiwo and Fawole [71] identified the effect of *Meloidogyne incognita* on cucumber in Nigeria bringing about remarkable yield reduction.

4.1.1 Root Knot disease

Root-knot is caused caused by *Meloidogyne* spp. All members of Cucurbitaceae family are susceptible. Typical symptoms observed on affected plants include stunted growth, pale green to yellow leaf colouration and wilting during the hot periods of the day [72]. This is due to reduced water uptake. Yield and quality of the fruit are greatly reduced. In heavy infections, plants will completely wilt and die as the nematode populations increase. When infected plants are removed from the soil, knobby, wart-like galls caused by the nematode can be seen singly or in clumps on the roots. Secondary infection by other soil organisms is common.

4.1.1.1 Management of Root knot disease

Measures that have been used to manage root knot disease include the use of chemically formulated nematicides, crop rotation, soil fumigation, practice of fallow period and using plant extract with nematicidal property.

5. Fungal diseases of cucumber

5.1 Genus: *Pseudoperonospora*

Members of the genus *Pseudoperonospora* are water moulds which include several species known for causing downy mildew infections on plants.

5.1.1 Downy mildew of Cucumber

Downy mildew is one of the important foliar diseases of members of *Cucurbitaceae* family. Unlike other cucurbits, cucumber is more susceptible to downy mildew [73]. *Pseudoperonospora* spp. is the causative pathogen of downy mildew on cucurbits generally [74]. The disease was reported on *C. sativus* L. in Egypt [75].

Serious losses result from downy mildew of cucumber under unfavourable environmental conditions. Temperature and humidity have been identified as important factors for disease infection and development [76]. The transmission of the fungus is dependent on the presence of infected cucurbit hosts in cultivated fields. Symptoms of downy mildew in cucurbits are almost exclusively confined to the leaves. In cucumber, characteristic symptoms are angular chlorotic lesion on foliage. The underside of leaf turns grey-brownish to purplish black with high humidity.

5.1.1.1 Management of Downy mildew

Planting should be planned to coincide with the time that environmental factors are less favourable to disease infestation. Crop should also be closely monitored for early identification of incidence of the disease. Adoption of good phytosanitary measures, use of healthy planting materials, chemical fungicides and bio-control agents are other effective means of managing the disease.

5.2 Genera: *Erysiphe, Sphaerotheca*

Two genera (Erysiphe, Sphaerotheca) have been identified to cause powdery mildews which affect economically important crops, herbs and woody species [77, 78]. Members of the genera are obligate plant pathogenic fungi. *Erysiphe* is the largest in the family Erysiphaceae, comprising about 873 species. About 806 plant species are found in the genus *Sphaerotheca*.

5.2.1 Powdery mildew of Cucumber

Powdery mildew of cucumber is caused by either *Erysiphe cichoracearum* or *Leveillula taurica* (Erysiphales) and *Sphaerotheca fuliginea* [76]. Powdery mildew had been identified on cucumber fields in Nigeria [79]. There are reports on the presence of *Leveillula taurica* on cucumber fields in Kenya, Libya, Morocco and

Senegal [80]. Powdery mildew infection is one of the easiest to spot because of the specificity of the symptoms. Transmission of the fungus occurs when conidia are dispersed over long distances by wind, through the movement or planting of infected cucurbits or by alternate hosts. Typical symptoms include white powdery growth in the upper leaf surfaces and stems of infected plants [81]. The lower or young leaves are most affected but mildew can be seen on any upper part of the plant. Infected plants are usually stunted and distorted. Fruits may also experience stunted growth.

5.2.1.1 Management of Powdery mildew

Planting resistant varieties, application of fungicides to delay infection and reduce disease incidence are common measures to manage powdery mildew. Additionally, use of non- hazardous biopesticide such as cinnamon oil, effective phytosanitation and regular weeding have also been recommended.

5.3 Genus: *Alternaria*

Species in the genus *Alternaria* have worldwide distribution because they are capable of existing in a variety of habitats, which explains their commonness and abundance. Members of the genus *Alternaria* are ascomycete fungi with major plant pathogens [82]. Lack of suitable hosts and unfavourable environment have been identified as some of the probable reasons for absence of *Alternaria* spp. in any environment [83]. The key distinguishing feature of the genus *Alternaria* is the production of dark coloured conidia with longitudinal and transverse septa. The spores are airborne and found in soil, water and on surfaces.

5.3.1 Alternaria leaf blight of Cucumber

Alternaria leaf blight of cucumber is caused *by Corynespora casiicola*. Although, cucumber is reported to be more susceptible to Alternaria leaf blight disease among the cucurbits, there are reports of wide host range distribution for the disease and evidences of infection abound in several other crops and weeds. Transmission of infection is carried out through wind over long distance and by splashing of water from diseased plants to previously unaffected ones over short distance. Lesions tend to appear first on older leaves in form of small circular spots. The spots are brown with light center which form dark concentric rings as they enlarge. Lesions formed on lower leaf surface tend not to be focused [84]. Fruit infection starts out as sunken brown spots and may later develop dark powdery appearance. The infection may also begin at the growing end while the entire fruit eventually turns brown. Spots on stems and petioles elongates more to distinguish the disease from other diseases such as angular leaf spot and anthracnose. In recent times, target spot has emerged with alarming severity and incidences on agriculturally important crops such as cotton (*Gossypium hirsutum*), soybean (*Glycine max*) and tomato (*Solanum lycopersicum*) in the Southeastern United States [85], causing losses of between 5–40%. In Africa, there seems to be scanty report of the disease on cucumber but it has been reported on such crops as tomato and tobacco in Nigeria [86], on potatoes in South Africa [87] and as a biocontrol agent of weed in Egypt [88]. *Corynespora casiicola* which was considered as a minor issue to production of major crops such as cotton, soybean is now causing epidemics in US by expanding its host range to previously non-host crops, it is therefore of importance considering it an emerging fungal disease which needs more attention in Africa.

5.3.1.1 Management of Alternaria disease of Cucumber

Alternaria disease of cucumber can be effectively managed by using resistant varieties for planting and practising crop rotation with non-cucurbitaceous crops. Also, destruction of volunteer alternate hosts, application of commercially recommended fungicides and adoption of good sanitation practices should be encouraged.

5.4 Genus: *Fusarium*

Members of the genus *Fusarium* are ubiquitous saprophytes which can be isolated from debris, roots, stems and seeds of a wide variety of plants [89]. *Fusarium oxysporum* is one of the important phytopathogen causing Fusarium wilt disease in more than a hundred species of plants [90]. *Fusarium* wilt of cucurbits is caused by *Fusarium oxysporum* f. sp. *niveum* which affects watermelon, *Fusarium oxysporum* f. sp. *melonis* which affects muskmelon and *Fusarium oxysporum* f. sp. *cucumerinum* which affects cucumber.

5.4.1 Fusarium wilt of Cucumber

Yield loss due to the infection of Cucumber Fusarium wilt disease has been estimated to be between 10–30%. Poor quality of harvested product also results causing huge economic losses [91]. *Fusarium* spp. cause wilt and root rot of affected plants. Invasion by the fungi starts from the root of the plant and progresses into the stems. Damping-off is the symptom of early infection which develops due to lower stem infections. Wilting generally starts on the older leaves and proceeds to the young ones. Wilt symptoms develop on one or few lateral vines in the beginning, while other branches remain apparently unaffected. However, under high inoculum conditions or in highly susceptible host species, the whole plant may wilt and die within a short time. Older plants may experience vascular browning, gummosis, and tylosis in xylem vessels [84]. Transmission of infection over short distances occur by the movement of infested soil, while the spread over long distances occur by using infected equipment and planting propagules. The disease may invade the fruits through the stem end.

5.4.1.1 Management of Fusarium wilt

Treatment of seed with chemical fungicides and planting resistant varieties are measures that can be used to manage the disease. Hot water treatment of seed at 52 °C for 30 minutes and crop rotation with distantly related crop is also helpful.

5.5 Genus: *Pythium*

Members of the genus *Pythium* are common worldwide [92]. Many species in the genus *Pythium* are important soil-borne pathogens which affect economically important crops such as cereals and vegetables [93]. Several other *P*ythium spp. are saprophytic, with some enhancing plant growth and displaying potential as a biocontrol agent [94]. About 8 *Pythium* species have been identified in Africa, *P. aphanidematum* being the commonest. Pythium-induced damping-off and root disease of cucurbits cause huge losses, sometimes as much as 100% [95].

5.5.1 Damping-off in cucumber

Damping-off caused by *Pythium* spp. constitute major constraint to cucumber production worldwide [96]. *Pythium* spp. were associated with wilt and root rot of hydroponically grown crops, including cucumber in South Africa [97]. The infection can occur before or after seeds germination. At pre-emergence, rotting of seed inside seed coat occurs. At post-emergence, damping-off appears as yellow to dark brown and water-soaked lesions are noticed on the root and hypocotyl tissue. With time, further root decay occurs, the hypocotyl shrivels and the seedling wilt. Inoculum of *Pythium* spp. can be transmitted from one place to another through airborne dust, soil, water, farm tools and equipment.

5.5.1.1 Management of Pythium spp

Chemical treatment, crop rotation, prevention of fruit contact with soil surface, reduction of fruit injury and biological control are being used to manage Pythium spp. Avoidance of overcrowding is also advisable.

6. Insect pest of cucumber

Three insect pests are of high importance to cucumber production. They are cucumber beetle (striped and spotted), green peach and melon aphids. Cucumber beetle cause direct feeding damage and can vector bacterial wilt. Aphids on the other hand vector viruses.

6.1 Striped and spotted cucumber beetles

Striped cucumber beetles, Acalymma vittata (Fabricius), and spotted cucumber beetles, *Diabrotica undecimpunctata howardi* (Barber) share a similar life cycle, and inflict similar damage to host plants. Both species of cucumber beetles not only feed on the roots, stems, foliage and fruit of cucumbers they also vector diseases of cucumber [98]. Cucumber beetles vector bacterial wilt and can transmit another important disease, however, spotted cucumber beetles do not vector bacterial wilt disease but can spread squash mosaic virus and there are reports that they can increase the incidence of powdery mildew, black rot, and Fusarium wilt [99]. The feeding injury and disease transmission takes place from emergence until they form runners. Beetle cause growth retardation and destruction of young seedlings causing loss to the farmers.

6.1.1 Management

There is usually a peak in beetle activity each spring that lasts two to four weeks. This is the most important time to control beetles. Monitoring is as important as control which should be done as soon as seedlings emerge because early treatment is essential for beetle management in cucumber fields. Treatment should be carried out at the peak of beetle activity with foliar insecticides to protect cucumber plants from beetle feeding and transmission of bacterial wilt.

6.2 Aphids (*Myzus persicae* and *Aphis gossypii*)

A number of aphid species which include green peach aphid (*Myzus persicae*) and melon aphid (*Aphis gossypii*) feed on cucumber and cause similar damage. The aphids suck plant fluids from stems, leaves, and other tender plant parts using their slender mouthpart to pierce. Downward curling and crinkling of the leaves of infested plants is part of the first sign of aphid damage. The aphids are often found on lower leaves, soft-growing tips, flower buds, and in some cases flowers. The feeding activity of the aphids usually causes a variety of symptoms, which can include reduced plant growth and vigour, mottling, yellowing, browning, curling, or wilting of leaves. All these can result in low yields and sometimes death of plant [100].

The Saliva that is injected into plant tissues by aphids can cause puckering and curling of leaves and such can protect them from natural enemies and help them evade substances applied for their control. Aphids feeding on flower buds and fruits can cause malformed flowers or fruits. Aphid is very prolific so populations can increase rapidly as each adult reproduces numerous nymphs in a very short period of time. It takes the green peach aphid just 10–12 days to complete one generation and can reproduce over 20 generations annually under mild climates [100]. Aphids generally are considered the most important vector for the transmission of viruses throughout the world with equal capability of both nymphs and adults [101].

6.2.1 Management

The first step to management is early identification. The ability of aphids have to multiply rapidly must be taken into account while monitoring for this pest. Plants should be checked on a regular basis, at least twice a week with special attention to the undersurface of the leaves in most cases problems occur toward the end of the growing season. There are a number of cultural and biological options available depending on environmental factors and available resources. Yellow sticky traps would be helpful in detecting aphids 2–3 weeks prior to planting and can also help in management afterwards. The use of a detergent and vegetable oil solution before destroying old crops can avoid winged virus-infected aphids from getting to nearby crops to overwinter. Chemicals both organic such as potassium soap and petroleum oil or primicarb. Endosulfan, Dimethoate, Lannate, Fulfil, and Actara and inorganic like Cypermethrin (0.01%), acetamiprid (0.01%), bifenthrin (0.01%) and malathion (0.05%) are used to control aphids.

Author details

Olawale Arogundade*, Titilayo Ajose, Joshua Olumide Matthew and Itunu Adeyemi Osijo
National Horticultural Research Institute, Ibadan, Oyo State, Nigeria

*Address all correspondence to: arogundade_olawale@yahoo.co.uk

References

[1] Wehner, TC. 2007. Cucumbers, watermelon, squash and other cucurbits. p. 474-479. In Encyclopedia of Food and Culture.

[2] Ene CO, Ogbonna PE, Agbo CU, Chukwudi UP. Studies of phenotypic and genotypic variation in sixteen cucumber genotypes. Chilean Journal of Agricultural Research. 2016;76 (3): 307-313

[3] FAO, 2006. Cucumber Production in Pakistan. http:// www.faostat.fao.org

[4] Eifediyi EK, Remison SU. Growth and yield of cucumber (*Cucumis sativus* L.) as influenced by farm yard manure and inorganic fertilizer. Journal of Plant Breeding and Crop Science. 2010; 216-220

[5] Umeh OA, Ojiako FO. Limitations of Cucumber (*Cucumis sativus* L.) Production for Nutrition Security in Southeast Nigeria. International Journal of Agriculture and Rural Development. 2018; 21(1):3437-3443.

[6] Owolabi AA. Yield and post-harvest qualities of cucumber as influenced by companion planting (project). Department of Crop Science and Horticulture: Federal University of Oye-Ekiti, 2017. 119pp

[7] Birch ANE, Begg GS. The future central role of IPM in EU crop protection: how can ecological research be put into practice? Proceedings Crop Protection in Northern Britain; 9-14 February 2010; pp. 317-333

[8] Sharma S, Kooner R, Arora R. Insect pests and Crop losses. In: Arora R., Sandbu S. (eds.), Breeding Insect Resistant Crops for sustainable agriculture. Singapore: Springer; 2017. 45-66.

[9] Ng JCK, Perry KL. Transmission of plant viruses by aphid vectors. Molecular Plant Pathology. 2004; 5: 505-511

[10] Cradock KR, da Graça JV, Laing MD. Viruses infecting cucurbits in KwaZuluNatal, South Africa. Revista Mexicana de Fitopatología. 2001; 19:251-252.

[11] Ibaba JD. Identification and characterization of viruses infecting cucurbits in the province of kwaZulu-Natal, republic of South Africa, with the purpose of developing transgenic virus-resistant cucurbits [thesis]. Department of Plant Pathology: University of KwaZulu-Natal, 2016. 136pp

[12] OEPP/EPPO (2005) Data sheets on quarantine pests. Cucumber vein yellowing ipomovirus. Bulletin OEPP/ EPPO Bulletin 35, 419-421

[13] Sastry KS. Plant virus and viroid diseases in the tropics. Introduction of Plant Viruses and Sub-Viral Agents, Classification, Assessment of Loss, Transmission and Diagnosis. New York: Springer; 2013. 11-97 in: Volume 1

[14] Daunde AT, Baghele RD, Khandare VS. Management of prevalent diseases of cucumber (*Cucumis sativus*) through integrated approach. International Journal of Current Microbiology and Applied Sciences. 2020;9(7): 2319-7706

[15] Al-Sadi AM, Al-Said FA, Al-Kaabi SM, Al-Quraini SM, Al-Mazroui SS, AlMahmooli IH, Deadman ML. Occurrence, characterization and of fruit rot of immature cucumbers under greenhouse conditions in Oman. Phytopathologia Mediterranea. 2011; 50(3):421-429.

[16] Agrios, GN. 2005. *Plant Pathology,* Academic Press, New York. 922 pp.

[17] Barth M, Hankinson TR, Zhuang H, Breidt, H. 2009. Microbiological

spoilage of fruits and vegetables. In Sperber, W. H. and Doyle, M. P. (Ed.), Compendium of the microbiological spoilage of foods and beverages, food microbiology and food safety, pp. 135-183. Springer Science-Business Medi, LLC.

[18] Mukhtar T, Kayani MZ, Hussain MA. Response of selected cucumber cultivars to *Meloidogyne incognita*. Crop Protection. 2012; 44:13-17

[19] Ayala-Doñas A, deCara-García M, Talavera-Rubia M, Verdejo-Lucas S. Management of soil-borne fungi and root-knot nematodes in cucurbits through breeding for resistance and grafting. Agronomy. 2020; (*10*) 1641.

[20] Gibbs A, Ohshima K. *Potyviruses* and the digital revolution. Annual Review of Phytopathology. 2010; 48: 205-223.

[21] Moury B, Verdin E. Viruses of pepper crops in the Mediterranean basin: A remarkable stasis. Advances in Virus Research. 2012; 86:127-162

[22] Lecoq H, Desbiez C. Virus of cucurbit crops in Mediterranean region: An ever-changing in: Loebenstein. G., Lecoq, H. (eds.), Viruses and virus diseases of vegetables in the Mediterranean basin. Advanced Virus Research. 84: 2012; 67-126

[23] Tsai W S, Abdourhamane IK, Knierim D, Wang JT, Kenyon L. First Report of *Zucchini yellow mosaic virus* Associated with Leaf Crinkle and Yellow Mosaic Diseases of Cucurbit Plants in Mali. Plant Disease. 2010; 94:923-923.

[24] Desbiez C, Lecoq H. *Zucchini yellow mosaic virus*. Plant Pathology. 1997; 46: 809-829

[25] Svoboda J, Leisova-Svobodova L, Amano, M. Evaluation of selected cucurbitaceous vegetables for resistance to *Zucchini yellow mosaic virus*. Plant Disease 2013; 97:1316-1321.

[26] Ravi S., Seema H., and Raj B. A Practical Guide to Identification and Control of Cucumber Diseases. TROPICA SEEDS PVT LTD, INDIA. 2014.

[27] Olczak-Woltman, H., Schollenberger M., Mądry, W. and Niemirowicz-Szczytt, K. Evaluation of cucumber (*Cucumis sativus*) cultivars grown in Eastern Europe and progress in breeding for resistance to angular leaf spot (Pseudomonas syringae pv. lachrymans). European Journal of Plant Pathology. 2008. 122:385-393.

[28] Miras M., Juárez M., and Aranda MA. Resistance to the emerging Moroccan watermelon mosaic virus (MWMV) in squash. Phytopathology. 2019. 1-33.

[29] Jackson, G. Pacific Pests and Pathogens - Mini Fact Sheet Edition, Cucumber downy mildew. http://www.pestnet.org/fact_sheets/mini/cucumber_downy_mildew_143.htm.

[30] Mohammed, H., Manglli, A., Zicca, S., El Hussein, A., Mohamed, M. and Tomassoli, L. First report of Papaya ringspot virus in pumpkin in Sudan. New Disease Reports. 2012. 26.

[31] Boubourakas IN, Avgelis AD, Kyriakopoulou PE, Katis NI. Occurrence of yellowing viruses (Beet pseudo-yellows virus, Cucurbit yellow stunting disorder virus and Cucurbit aphid-borne yellows virus) affecting cucurbits in Greece. Plant Pathology. 2006. 55: 276-283.

[32] Deadman, MI. Pythium damping off and root-rot. Pythium damping off and root-rot, in Compendium of Cucurbit Diseases and Pests. American Phytopathological Society (2017).

[33] Parry JN, Persley DM. Carrot as a natural host of *Watermelon mosaic virus*.

Australasian Plant Pathology. 2005; 34: 283-284.

[34] Ali A, Natsuaki T, Okuda S. The complete nucleotide sequence of a Pakistani isolate of *Watermelon mosaic virus* provides further insights into the taxonomic status in the *Bean common mosaic virus* subgroup. Virus Genes. 2006; 32: 307-311

[35] Ayo-John EI, Olorunmaiye PM, Odedara OO, Dada OB, Abiola KO, Oladokun JO. Assessment of Field-Grown Cucurbit Crops and Weeds within Farms in South-West Nigeria for Viral Diseases. Notulae Scientia Biologicae. 2014; 6(3):321-325

[36] Sydänmetsä M, Mbanzibwa DR. Occurrence of *Cucumber mosaic virus*, *Zucchini yellow mosaic virus* and *Watermelon mosaic virus* in cultivated and wild cucurbits in the coastal areas of Tanzania. African Journal of Agricultural Research. 2016; 11 (40):4062-4069

[37] Laney AG, Avanzato MV, Tzanetakis IE. High incidence of seed transmission of *Papaya ringspot virus* and *Watermelon mosaic virus*, two viruses newly identified in *Robinia pseudoacacia*. European Journal of Plant Pathology. 2012; 134: 227-230.

[38] Finetti-Sialer MM, Mascia T, Cillo F, Vovlas C, Gallitelli, D. Biological and molecular characterization of a recombinant isolate of *Watermelon mosaic virus* associated with a watermelon necrotic disease in Italy. European Journal of Plant Pathology. 2012;132: 317-322.

[39] Arocha Y, Vigheri N, Nkoy-Florent B, Bakwanamaha K, Bolomphety B, Kasongo M, Betts P, Monger W. A, Harju V, Mumford RA, Jones P. First report of the identification of *Moroccan watermelon mosaic virus* in papaya in Democratic Republic of Congo. Plant Pathology. 2008; 57: 387-387

[40] McKern NM, Strike PM, Barnett OW, Ward CW, Shukla, DD. *Watermelon mosaic virus-Morocco* is a distinct *Potyvirus*. Archives of Virology. 1993;131: 467-473

[41] Lecoq H, Dafalla G, Desbiez C, Wipf-Scheibel C, Delecolle B, Lanina T, Ullah Z, Grumet R. Biological and molecular characterization of *Moroccan watermelon mosaic virus* and a potyvirus isolate from Eastern Sudan. Plant Disease. 2001; 85:547-552.

[42] Owolabi AT, Rabenstein F, Ehrig F, Maiss Edgar M, Vetten, HT. Strains of *Moroccan watermelon mosaic virus* isolated from *Lagenaria brevilflorus* and *Coccinia barteri* in Calabar, South-eastern Nigeria. International Journal of Virology. 2012; 8:258-270.

[43] Yakoubi S, Desbiez C, Fakhfakh H, Wipf-Scheibel C, Marrakchi M, Lecoq, H. Biological characterization and complete nucleotide sequence of a Tunisian isolate of *Moroccan watermelon mosaic virus*. Archives of Virology. 2008b; 153: 117-125

[44] Cherif C, Ezzaier K, 1987. Viruses of cucurbits in Tunisia. In: Proceedings of the Seventh Conference of the Mediterranean Phytopathological Union, 1987. Granada, Spain, 137-138

[45] Omar AF, El-Kewey SA, Sidaros SA, Shimaa AK. Egyptian isolates of *Papaya ringspot virus* form a molecularly distinct clade. Journal of Plant Pathology. 2011; 93:1224. [http://dx.doi.org/10.4454/jpp.v93/3.1224]

[46] Mohammed H, Manglli A, Zicca S, El Hussein A, Mohamed M, Tomassoli L. First report of *Papaya ringspot virus* in pumpkin in Sudan. New Disease Reports. 2012; 26:26.

[47] Ibaba JD, Laing MD, Gubba A. Incidence and phylogeny of viruses infecting cucurbit crops in KwaZulu-Natal, Republic of South Africa. Crop Protection. 2015; 75:46-54

[48] Palukaitis P, García-Arenal F. *Cucumoviruses*. Advanced Virus Research. 2003; 62:241-323.

[49] Bujarski J, Figlerowicz M, Gallitelli D, Roossinck MJ, Scott SW. *Bromoviridae*. In: Virus taxonomy: classification and nomenclature of viruses: Ninth Report of the International Committee on Taxonomy of Viruses. Ed: King, A.M.Q., Adams, M.J., Carstens, E.B. and Lefkowitz, E.J. San Diego: Elsevier; 2012. 935-946.

[50] Jacquemond M. 2012. *Cucumber mosaic virus*. In: Loebenstein, G. & Lecoq, H. (ed.), Viruses and virus diseases of vegetables in the Mediterranean basin. Advances In virus Research vol. 84. USA: Elsevier; 2012. 439-504

[51] Kumari R, Bhardwaj P, Singh L, Zaidi AA, Hallan V. Biological and molecular characterization of *Cucumber mosaic virus* subgroup II isolate causing severe mosaic in cucumber. Indian Journal of Virology. 2013 24: 27-34.

[52] Palukaitis P, Roossinck MJ, Dietzgen RG, Francki RIB. *Cucumber mosaic virus*. Advanced Virus Research. 1992; 41:281-349.

[53] Arogundade O, Ajose T, Osijo I, Onyeanusi H, Matthew J, Aliyu TH. Management of Viruses and Viral Diseases of Pepper (Capsicum spp.) in Africa: Capsicum. DOI: 10.5772/intechopen.92266

[54] Mauck, KE, De Moraes, CM, Mescher, MC. 2010 Deceptive chemical signals induced by a plant virus attract insect vectors to inferior hosts. Proceedings of the National Academy of Sciences 107:3600-3605.

[55] Mochizuki T, Ohki ST. *Cucumber mosaic virus*: viral genes as virulence determinants. Molecular Plant Pathology. 2012; 13: 217-225

[56] Arogundade O, Balogun OS, Kumar PL. Seed transmissibility of *Cucumber mosaic virus* in Capsicum species. International Journal of Vegetable Science. 2019; 25(2): 146-153. [DOI:10.1080/19315260.2018.1487498]

[57] Martelli GP, Agranovsky AA, Bar-Joseph M, Boscia D, Candresse T, Coutts RHA, Dolja VV, Hu JS, Jelkmann W, Karasev AV, Martin RR, Minafra A, Namba S, Vetten HJ. 2012. In: Virus taxonomy: classification and nomenclature of viruses: Family Closteroviridae. Ninth Report of the International Committee on Taxonomy of Viruses. Ed: King, A.M.Q., Adams, M.J., Carstens, E.B. and Lefkowitz, E.J. San Diego: Elsevier; 2012. 987-1001

[58] Liu HY, Duffus, JE. *Beet pseudo-yellows virus* - purification and serology. Phytopathology. 1990; 80: 866-869.

[59] Kado CI. Erwinia and related Genera. In: Dworkin, M., Falkow, S., Rosenberg, E., Schleifer, KH., Stackebrandt, E. (eds.). The Prokaryotes.

[60] Shapiro LR, Seidl-Adams I, De Moraes CM, Stephenson AG, Mescher MC. Dynamics of short and long term association between a bacterial plant pathogen and its arthropod vector. Scientific Reports. 2014;4: 4155

[61] Bradbury JF. *Erwinia traceiphila*. Description of Fungi and Bacteria. IMI Descriptions of Fungi and Bacteria. 1970; 24.

[62] Mohammadi AS. Managing Bacterial wilt caused by Erwinia tracheiphila on muskmelon with early control of striped cucumber beetle (*Acalymma vittatum* F.) and through

varietal selection [thesis]. Purdue University, 2016. 175pp

[63] Morris CE, Sands DC, Vinatzer BA, Glaux C, Guilbaud C, Buffiere A, Yan S, Dominguez H, Thompson BM. The life history of the plant pathogen Pseudomonas syringae is linked to the water cycle. International Society for Microbial Ecology Journal. 2008; 2:321-334.

[64] Bhat NA, Bhat KA, Zargar MY, Teli MA, Nazir M, Zargar SM. Current Status of Angular Leaf Spot (*Pseudomonas syringae* pv. *lachrymans*) of Cucumber: A Review

[65] Pohronezny K, Larsen PO, Emmatty DA, Farley JD. Field studies of yield losses in pickling cucumber due to angular leaf spot. Plant Disease Reporter. 1977;61: 386-390.

[66] El-Sadek SAM, Abdel-Latif MR, Abdel-Gawad TI, Hussein NA. Occurrence of angular leaf spot disease in greenhouse cucumbers in Egypt. Egyptian Journal of Chemistry. 1992; 27: 157-175.

[67] Mohamed ZK, El-Hindawy HH, Fayed OS. Physiological and biochemical studies on phytopathogenic bacteria isolated from cucumber in Egypt. Egyptian Journal of Microbiology 2000; 35: 1-20.

[68] Archidona-Yuste A, Cantalapiedra-Navarrete C, Liebanas G, Rapoport HF, Castillo P, Palomares-Rius JE. Diversity of root-knot nematodes of the genus *Meloidogynidae* associated with olive plants and environmental cues regarding their distribution in southern Spain 2018

[69] De Waele D, Elsen A. Challenges in tropical plant nematology. Annual Review of Phytopathology. 2007; 45: 457-485.

[70] Omar NH, Adam MAM. Survey of plant parasitic nematode associated with cucumber plant (*Cucumis sativus* L.) in Al-Jabal Al-akther region. Al-Mukhtar Journal of Sciences. 2018; 33 (1): 36-44.

[71] Aminu- Taiwo BR, Fawole B. Pathogenicity of Meloidogyne incognita on four cucumber cultivars (*Cucumis sativus* L.). Inaugural conference of Nigeria society of Nematologist (Book of Abstract) 16-17th October, 2012.

[72] Noling JW. Nematode Management in Cucurbits (Cucumber, Melons, Squash). UF/IFASExtension. 2017.

[73] Ojiambo, PS, Paul PA, Holmes GJ. A quantitative review of fungicide efficacy for managing downy mildew in cucurbits. Phytopathology. 2010;100(10): 1066-1076.

[74] Lebeda, A. and Cohen, Y. Cucurbit downy mildew (*Pseudoperonospora cubensis*) biology, ecology, epidemiology, host-pathogen interaction and control. European Journal of Plant Pathology. 2011; 129:157-192.

[75] Mahrous MM, Fadi FA, Youssef KS. Downey of cucumber: a potent problem in Egypt. First National Conference of pests and diseases of vegetables in Egypt. Ismailiae. 1985;1070-1084

[76] FAO 2007. Cucumber Integrated Pest Management: An Ecological Guide. FAO Inter-Country Program for the Development and Application of Integrated Pest Management in Vegetable Growing in South and South East Asia; 2007.88pp

[77] Braun U, Cook RTA. 2012 – Taxonomic manual of the Erysiphales (powdery mildews). CBS Biodiversity Series No. 11. CBS–KNAW Fungal Biodiversity Centre, Utrecht.

[78] Abasova LV, Aghayeva DN, Takamatsu S. Notes on the genus Podosphaera Erysiphales) on rosaceous hosts in Azerbaijan. Proceedings of

the Institute of Botany ANAS. 2017; 37:14-19

[79] Labo AU, Khan AA, Ahmad G. Incidence of powdery mildew on cucurbit plants in Katsina, Nigeria. Journal of Phytology. 2019; 11: 5-9

[80] Farr DF, Rossman AY. Fungal databases System. Mycol. Microbiology lab., online publication. 2018 ARS USDA

[81] Chand G, Kumar S, Kureel RS, Singh D. Diseases of cucurbits, In: Diseases of vegetable crops: Diagnosis and management (Eds Singh, D., Chowdappa, P and Sharma, P.) Today & Tomorrows's Printers and Publishers, New Delhi. 2014; p 185-242.

[82] Lawrence DP, Park MS, Pryor BM. Nimbya and Embellisia revisited, with nov. comb for Alternaria celosiae and *A. perpunctulata*. Mycological Progress. 2012 11: 799-815.

[83] Rotem J. The genus *Alternaria*. Biology, epidemiology, and pathogenicity. American Phytopathological Society Press, St. Paul, MN.1994

[84] Sharma A, Katoch V, Rana C. Important Diseases of Cucurbitaceous Crops and their Management; 2016. 24pp

[85] Sumabat LG, Kemerait RC, Jr and Brewer MT. Phylogenetic Diversity and Host Specialization of *Corynespora casiicola* responsible for emerging target spot disease of Cotton and other crops in the Southeastern US. Phytopathology. 2018; 108:892-901

[86] Fajola AO, Alasoadura S.O. 1973. Corynespora leaf spot, a new disease of tobacco (Nicoti-ana tabacum). Plant Disease Reporter 57, 375-8

[87] Dube JP. Characterisation of *Alternaria alternata* isolates causing brown spot of potatoes in South Africa (thesis). Department of Microbiology and Plant Pathology: University of Pretoria, 2014, 108pp

[88] Sanodiya BS, Thakur GS, Baghel, RK, Pandey AK, Prasad, GBK, Bisen PKS. Isolation and characterization of tenuazonic acid produced by *Alternaria alternata*, a potential bioherbicidal agent for control of *Lantana camara*. Journal of Plant Protection Research. 2010; 50:133-139.

[89] Leslie J, Summerell BA. The Fusarium Laboratory Manual. In: Practical Approaches to Identification. Blackwell. 2006; 11:101-110.

[90] Liu B, Zhu YQ, Zhou HT, Zhang SQ, Xie GL, Zhang SS. Advances in crop wilt disease. Journal of Xiamen University (in Chinese). 2004; 43:47-58.

[91] Li J, Yang Q, Zhang SM, Wang YX, Zhao XY. Evaluation of biocontrol efficiency and security of a *Bacillus subtilis* strain B29 against cucumber Fusarium wilt in field. China Vegetables. 2009; 2:30-33.

[92] Dick MW. 1990. Keys to Pythium. University of Reading, Reading, UK.

[93] Matoba Y, Kondo N, Akino S, Kodama F, Naito S, Ebe, S. Identification and pathogenicity of *Pythium* species causing damping-off of kidney bean. Journal of General Plant Pathology. 2008; 74: 81-85

[94] Higginbotham RW, Paulitz TC, Kidwell KK. Virulence of *Pythium* species isolated from wheat fields in eastern Washington. Plant Disease. 2004; 88:1021-1026.

[95] Al-Sadi AM, Deadman ML, Al-Said FA, Khan I. First Report of *Pythium splendens* associated with severe wilt of Muskmelon (*Cucumis melo*) in Oman. Plant Disease. 2008; 92:2

[96] Halo BA, Al-Yahyai RA, Mahara Chchikumbura SSN, Al-Sadi AM.

Talaromyces variabilis interferes with *Pythium aphanidermatum* growth and suppresses Pythium-induced damping-off of cucumbers and tomatoes. Scientific Reports. 2019; 9:11255.

[97] Gull C, Labuschagne N, Botha WJ. *Pythium* species associated with Wilt and root rot of Ojiambo, P. S., Paul, P. A. and Holmes, G. J. African Plant Protection. 2004; 10:2 (2010). A quantitative review of fungicide efficacy for managing downy mildew in cucurbits. Phytopathol. 100

[98] Akhilesh S, Chanchal R, Kumari S. Insect Pests of Cucurbits and Their Management. In book: Handbook of Cucurbits: Growth, Cultural Practices and Physiology Edition: First Chapter: 19Publisher: CRC Press 2016 Editors: Mohammad Pessarakli

[99] Diver S, Hinman T. Management factsheet, Cucumber beetles: organic and biorational integrated pest management. The National Center for Appropriate Technology (NCAT) Butte USA.

[100] Capinera, JL. Melon Worm. Diaphania hyalinata. Featured Creatures. Division of Plant Industry, Department of Entomology and Nematology Florida Cooperative Extension Service, Institute of Food and Agricultural Sciences, University of Florida, Gainesville, FL. Publication # EENY-163. http://edis.ifas.ufl.edu/in320. 2005.

[101] Barbercheck, ME. Biology and management of aphids in organic cucurbit production systems. www.extension.org/pages/60000/biology-and-management-of-aphids-in organic-cucurbit-production-systems. 2014

Section 6

Breeding Progress

Chapter 13

Classical Genetics and Traditional Breeding in Cucumber (*Cucumis sativus* L.)

Gograj Singh Jat, Tusar Kanti Behera, Suman Lata and Sachin Kumar

Abstract

Rapid progress has been made in classical genetics and traditional breeding in cucumber for various quantitative and qualitative traits which greatly benefited the development of superior varieties suitable for open field and protected cultivation. The different breeding methods like plant introductions, hybridizations, pedigree selection, recombination breeding and marker assisted selection have been employed successfully in developing superior varieties and F_1 hybrids. The development of new varieties with earliness, high-yield and resistance to diseases (powdery mildew, downy mildew and tolerant to virus) through selection of superior parental lines followed by hybridization and marker assisted introgression of desired genes was a game changer in cucumber breeding. The exploitation of gynoecious sex along with parthenocarpic traits in traditional breeding has made revolution in cucumber cultivation throughout the world which enabled the adoption of cucumber crop by farming community on large scale. Molecular markers technology could be exploited to overcome the obstacle of traditional breeding by accelerating the breeding cycle and selection of desirable traits. The high density genetic maps for various traits have been constructed in cucumber to detected quantitative traits loci (QTLs) for genetic enhancement in different market classes of cucumber. Therefore, this chapter highlighted the concepts of genetic foundations for advancement made in cucumber breeding.

Keywords: cucumber, traditional breeding, classical genetics, molecular breeding

1. Introduction

Cucumber (*Cucumis sativus* L) is an important vine species of the Cucurbitaceae family. Cucumber is the fourth most important vegetable crop worldwide and is a model system for other Cucurbitaceae family which is used for studying several significant biological processes [1]. Cucumber is originated from India, particularly southern foot-hills of Himalayan region. It was domesticated in India from its wild relative, *Cucumis sativus* var. *hardwickii* 3000 years ago [2, 3]. It is commercially grown in the tropical and subtropical regions of the world. The fruits are widely consumed as salad at immature stage. Cucumber is high in water content and low in calories, fat, cholesterol, and sodium [4] and good source of mineral nutrients (Ca, Mg, P & K) and medicinal properties such as antioxidant, anti-inflammatory, and anti-cancer benefits. Cucumbers are also used for digestive benefits and mood stability when modulating stress. Cucumbers fortify cells so they may retain hydrated and work at the highest levels, and may slow age-related cellular deteriorations [5]. The available genetic diversity within the cultivated cucumber is very low [2, 6–9] which is the major impediment in the genetic improvement of various cucumber market classes [10]. Thus, increasing the genetic diversity of cultivated cucumber is an important task for public sector research [11]. The cultivated cucumber has narrow genetic base with 3–8% polymorphism within the cultivated genotypes, and 10–25% between botanical varieties [12]. Earliness, high yield, uniform fruit shape, size, color and better quality are prerequisites for the release of the cucumber varieties and F_1 hybrids for open field condition. In addition to these characters, gynoecious and parthenocarpic traits are desirable for green house cucumber production. Identification of genotypes tolerance to drought is also one of important breeding objective in cucumber [13]. Cucumber is a monoecious vegetable crops species. However, several gynoecious varieties and F_1 hybrids have been developed by introgression of *F* locus (gynoecious gene) in the background of different market classes of the cucumber for commercial production. The utilization of gynoecious lines are economical and easier for hybrid seed production by reducing the cost of male flower pinching and hand pollination [14]. The present day cucumber F_1 hybrids for open field production derived from the cross between gynoecious × monoecious and monoecious × monoecious where as green house grown F_1 hybrids are the result of the cross of gynoecious × gynoecious lines with parthenocarpic traits. The evaluation and selection of the genotypes only based on phenotype characters for high yield and stability of gynoecious sex form require many years in multiple environments which is very expensive and time consuming process. The whole genome sequencing in cucumber [15] have opened the way to utilize the DNA markers viz. simple sequence repeats (SSRs) for gene mapping, marker assisted selection and marker trait association for several economic traits in cucumber [16–18]. With the development of high density linkage map several traits have been mapped in cucumber including, flowering time [19] fruit quality [20], diseases [17, 21], yield [22], fruit spines [23], fruit color [8], chromosomal mapping and QTL analysis of resistance to downy mildew [24–26], yellow fruit flesh [27], pleiotropic andromonoecy and carpel number [28, 29]. QTLs have been identified in Sikkim cucumber (*Cucumis sativus* var. *sikkimensis*) for important horticultural traits including flowering time, fruit size, flesh thickness, fruit spines, fruit color [30].

2. Classical genetics in cucumber

2.1 Genetics of quantitative and qualitative characters

The deep understanding about cucumber crops biology was only possible due to mendelian's classical genetics which have made possible for the cucumber breeders to develop improved varieties and F_1 hybrids. The knowledge about different genes which affects economic traits facilitated breeders to develop proper genetic resources for the development of trait specific genetic stock for further use for genetic improvement of cucumber. For example, the size of the population will be much smaller if a breeder is selecting for a trait controlled by a single gene, than if the trait is controlled by multiple genes with a large environmental influence. The application of Mendelian genetics using classical techniques in cucumber has facilitated the discovery of a number of genes for yield, quality, plant architecture, and disease and pest resistance in both slicing and pickling cucumber. The yield in cucumber is enhanced by utilizing gynoecious as one of the parent in breeding programme which promote higher female to male sex ratio. Sex ratio (female: male), fruit weight and fruit size are the direct yield components in cucumber breeding.

2.2 Sex expression

In commercial cucumber cultivars, the type of sex forms (gynoecious or monoecious) and the amount of their expression is important because these have a direct effect on harvesting date, production and productivity of this crop. Flowering time in cucumber played a critical role in fetching early market price and increase in fruit yield for the growers. In cucumber, the type of sex from (gynoecious or monoecious) and the amount of their expression have direct effect on harvesting time, production and productivity. The sex expression also played a vital role in seed production as well as development of new plant types. The flowering traits like node number at which first female flower appear, days to first pistillate flower opening, and male: female (♂:♀) flower ratio (sex ratio) are the important traits for determination of earliness and fruit yield. The sex expression in cucumber is controlled by three genes, *F*, *M*, and *A* [31]. The degree of female flower expression is controlled by *F/f* gene [31, 32]. The *F* locus determines the amount of femaleness (*FF* > *Ff* > *ff*). Gynoecious sex expression in F_1 hybrid of cross of gynoecious × monoecious is governed by partial dominance ([33]; Perl-Treves and Rajagopalan [34] where as in gynoecious × subandroecious, it was governed by multiple genes [35]. single gene with dominant or incomplete dominance [36], single dominant gene [8, 37–39] and oligogene with some modified genes [40], three major QTLs conferring subgynoecy in cucumbers [41]. Seven gynoecious QTLs were detected on chromosomes 5 and 6 in backcross population [42]. These studies suggested that gynoecious is an important economic trait for determinant of earliness and yield in cucumber.

2.3 Parthenocarpy

The discovery of parthenocarpy in cucumber has led to the development of seedless fruit in combination with gynoecious trait. Gynoecy coupled with parthenocarpy is a yield and quality related parameter and a high value vegetable crop suited for protected cultivation because these varieties do not require pollination for fruit setting. The fruits of greenhouse parthenocarpic cucumber varieties are also mild in flavor, seedless and have a thin skin that does not require peeling. Still the genetics of parthenocarpy is not well understood in cucumber which is utmost important for efficient breeding procedure. Pike and Peterson [43]; Kim et al. [44] and Jat et al. [38] suggested that an incomplete dominant gene is responsible for parthenocarpic fruit development. Single recessive gene is responsible for parthenocarpy in cucumber [45]. The growing environmental conditions and epistatic interactions influence the parthenocarpy trait [46, 47] and two additive dominant epistatic major genes and additive dominant polygenes [35]. Seven QTLs for parthenocarpy were detected on chromosome 5 and 7 (*parth5.1* and *parth7.1*) and two on chromosome 6 (*parth6.1* and *parth6.2*) [48]. One major effect QTL (parth 2.1) was identified controlling parthenocarpy [49, 50]. The identification of QTLs is a valuable resource for cucumber breeders for development of parthenocarpic cultivars.

2.4 Fruit characters

The improvement in fruit traits including shape, size, and color is an important target in cucumber breeding. The fruit traits like fruit weight, length diameter and number of fruit per plant are directly related to yield. The other traits like shape index (length: diameter and length/stalk ratio), fruit skin (size of spines, dull or uniform fruit color) and fruit bitterness decides the market value of fruits which are important particularly for slicing cucumber [51]. A single recessive gene controls fragrance in cucumber [52], fruit color traits were controlled by two major genes, single recessive gene, (*w*), was identified that controls white immature fruit color [53], single recessive gene controlled the inheritance of quantity of beta-carotene [54]. Green flesh color is one of the most important and commercial trait of cucumber fruit was controlled by a major effect QTL [55]. The bent characters of the fruit were quantitative inheritance controlled by multiple genes and major genes [56]. Twin fused fruits were controlled by single recessive gene [57]. Two loci controlling fruit bitterness in cucumber [24–26]. The QTLs were also identified in cucumber for several economic traits including fruit spine, skin colors, fruit netting, fruit size, hollow size, flesh thickness variation [58], fruit carpel number, sex expression, fruit length, fruit diameter, fruit shape, fruit number, and powdery mildew resistance [28, 29].

2.5 Inheritance for resistance to diseases

Several diseases infect cucumber which results in huge yield losses in production. The common disease infecting cucumber is powdery mildew, downy mildew, anthracnose, scab, angular leaf spot and virus diseases (cucumber mosaic virus (CMV), papaya ring spot virus, (PRSV), zucchini yellows mosaic virus). The management of these diseases increase the cost of production because of increasing use of chemicals for control of diseases which also has adverse implication on environment and ecology. The development of resistant sources and breeding of resistant varieties are economical, healthier and eco-friendly approach. There are different approaches for breeding resistance against different kinds of stresses which involve both conventional and modern tools. Achieving the goal of development of a resistant variety also needs attention on the durability of resistance for which the knowledge of resistance on the inheritance, expression and interaction with related genes and environment aspects are required.

2.6 Powdery mildew

It is one of the most severe foliar diseases in cucumbers but the inheritance of its resistance still remains unclear which is the major bottle-neck in the development of improved cucumber cultivars. Several fungicides are available to control this disease but their application increases environmental burdens and financial pressures on growers [59]. Therefore, development of the suitable genotypes resistance to powdery mildew is need of the hours. Several resistant cucumber cultivars have been identified, are only resistant to powdery mildew, such as PI 250147 [60], S06 [61], and SSSL0.7 [62]. The resistance in the Puerto Rico 37 variety was controlled by multiple recessive genes [63], PM resistance in the PI 197087 variety was controlled by 1–2 major and 1–2 minor genes [64], one and two recessive genes in PI 2008151 and Natsufushinari, respectively [65], major recessive gene (s) and a dominant gene (R) and a dominant suppressor gene (I) in P1212233 and P123514, [66, 67]. The genetic loci conferring resistance to PM have been gradually identified *pm-1, pm-2, pm-3* and *pm-h* loci [60]. A dominantly inherited PM resistance gene, *Pm1.1* present in Chinese Long line Jin5–508, which has been mapped. Most resistant cucumber cultivars become susceptible to PM at low temperatures. Collectively, the inheritance of cucumber PM resistance is complex and affected by temperature and region.

2.7 Downy mildew

Understanding the inheritance of downy mildew is fundamental to successful cucumber breeding programs. Downy mildew resistance is controlled by a single recessive gene [68, 69], pair of dominant and recessive interacting genes [70], two recessive genes in PI 197088 and a single recessive gene in 'Poinsett' [71], three recessive genes, *dm-1*, *dm-2* and *dm-3* [72], two incompletely dominant genes [73], inheritance for downey mildew resistance is quantitative [24–26, 74], number of genes, dominant, partially dominant, or recessive are responsible for resistance [75], multiple recessive genes [76]. The cucumber accession PI 197087 from India and its derivatives like Gy14 are resistant to downy mildew that is conferred by *dm-1.* The *dm-1* conferred DM resistant was less effective since 2004 when new DM pathogen strains emerged. The different results for inheritance of downy mildew resistance may be due to some factors like pathogen variability, environmental factors (temperature, relative humidity, inoculum movement etc.), mechanism of resistance and source of resistance. Recently, two major-effect QTLs for resistance to DM strain (*dm4.1* and *dm5.2*) were identified from PI 197088 and PI 330628.

2.8 Virus diseases

Chemical treatments against cucumber mosaic virus (CMV) are not effective; therefore genetic resistance is considered the primary line of defense. Genetic analysis revealed that CMV resistance in cucumber is quantitatively inherited [77] where as in *Cucumis sativus* var. *hardwickii* (wild relative) is controlled by single recessive gene [78]. The inheritance of resistance to zucchini yellow mosaic virus in cucumber is controlled by single recessive gene (). The genes resistant to viruses has been mapped on chromosome 6 including *psrv* for PSRV, *wmv* for WMV and *cmv6.1* for CMV [77, 79, 80]. Recently, cucurbit aphid-borne yellows virus (CABYV) has been reported in cucumber from India [81].

3. Traditional breeding

3.1 Traditional breeding objectives and achievements

The development of high yielding cultivars with better fruit quality is the major goal of cucumber breeding programs across the globe. Due to the introgression of gynoecious sex form in the background of different cucumber market classes, the development of varieties with earliness characters along with high female: male flowers ratio is also an important objective for cucumber breeding. Therefore, rapid advancement has been made in the development of gynoecious and parthenocarpic cucumber lines/varieties suitable for green house cultivation. Cucumber breeders have developed several genetic stocks for pest and disease resistance. The introgression of pest and disease resistance gene from available sources should be the prime goal for most of the cucumber breeding programmes in future for the development of multiple disease resistance (MDR) varieties including virus resistance. The various fruit quality traits have been selected in cucumber depending on regional preference of the consumers. These quality traits include fruit size, shape (length and diameter), color (light green, dark green), shelf life, seediness and nutritional quality. Cucumber breeders have combined several sex expression and fruit quality genes to develop desirable genotypes. Many pickling cucumber hybrids were developed by crossing gynoecious lines with monoecious lines. These hybrids are predominantly gynoecious in nature which provides pollens for fruit set which enhances total yield in pickling cucumber. Breeding technique for achieving these goals differ among breeders as does his/her definition of quality fruits. Gynoecious sex has also been combined with parthenocary trait to set fruit without pollen particularly for green house cultivation, since it eliminates the need for pollination and produces superior quality fruits. Gynoecious sex expression has been commercially exploited for the hybrid seed production since the female plants do not produce male flowers. Recently, the cucumber breeders have been interested in the development of carotene rich (yellow cucumber) lines. Several cucumber varieties have been developed for early characters, high yield, and resistance to diseases using heterosis breeding. The most commonly used breeding methods for the improvement of cucumber is inbreeding, single plant selection from segregating populations, heterosis breeding and presently marker assisted selection (MAS). The most common breeding procedure followed in cucumber is the selection from a local cultivar. Several F_1 hybrids have been developed in cucumber both from public and private sector through hybridization across the globe. Cucumber is easily grown, indeterminate plant types, offer plenty of large flowers to work with over a long period of time where a large variation is observed for quantitative and qualitative characters. The use of gynoecious lines as one of the parent in breeding programme has made possible to enhance the area under hybrid cultivation. Recent QTL mapping and cloning studies in cucumbers for many quantitative traits will present a complete picture on the genetic architecture of these traits. The identification of molecular markers (SNPs and SSRs) for earliness, yield and quality traits would be directly useful in genome selection to expedite cucumber breeding in different market classes. The cloning of QTLs and genome wide association studies (GWAS) for important quantitative traits will be more innovative and future-focused for trait specific breeding of cucumber.

3.2 Breeding methods

Several breeding methods have been employed for the genetic improvement of cucumber depending on the specific breeding objectives. Single plant selection, single seed descent method, mass selection, simple backcross breeding, pedigree selection, hybridization, use of sex inheritance and chemicals in breeding, and population improvement and extraction of inbred lines are the most common methods used. In recent time, marker assisted selection derived lines have been developed for the improvement of quantitative and qualitative traits in different cucumber market classes. Simple backcross breeding is quite useful for transferring characters governed by single genes e.g. disease resistance or quality traits from donor lines to more stable recurrent parents. Often, six generations of selection and backcrossing to the recurrent parent are required to recover the desired genotypes (recurrent parent + additional trait). Single seed descent method is useful for the development of inbred lines by self pollination. In chemical breeding the gynoecious lines are treated with Silver nitrate/silver thiosulphate to induce hermaphrodite flowers for pollination. The population improvement method is based on recurrent selection and aims at long term gains for the characters having low to moderate heritability. The exploitation of hybrid vigor in cucumber is desirable due to high heterosis for earliness, yield and disease resistance. Heterosis breeding can exploit the genetic diversity present in cucumber for various growth and yield characters. In Western countries almost 90% of the area of cucumber is under F_1 hybrids.

3.3 Heterosis breeding

Cucumber is a monoecious and cross pollinated crop, there is a great scope for exploitation of heterosis. High level of hybrid vigor could be obtained with the involvement of diverse parents. Several studies have been conducted to identify the best heterotic combination for earliness, yield and quality traits in cucumber. Appreciable heterosis was observed over better parent and top parent for many economic traits like node number of first female flowers, number of fruits per plant, days to fruit set, days to first fruit harvest, yield per plant [82–89]. Significant heterosis has also been reported for earliness characters using gynoecious lines [82, 83, 90] and quality characters [88]. The hybrid combinations of gynoecious × gynoecious and gynoecious × monoecious showed maximum heterosis followed by monoecious × monoecious hybrids for earliness and yield per plant [83, 90]. Therefore yield of the cucumber can be enhanced by using gynoecious line as one of the parent in future breeding programme. Hence breeder should concentrate mainly number of fruits rather than the fruit size to increase yield in cucumber.

3.4 Limitations of traditional breeding

Traditional breeding has the major limitations of its dependency only on selection of traits based on morphological markers (leaf characters, flowering traits including sex ratio, fruit color, fruit size, shape etc.) from a segregating population. Traditional breeding has been effectively utilized for the improvement of qualitative characters. Traits like beta carotene containing cucumbers (linked to orange flesh) and parthenocarpy (linked to seedless) are the classical examples of selection using morphological markers. Traditional breeding has also been exploited for major changes in important quantitative characters including early maturity, fruit size, free from bitterness and fruit yield. Further genetic improvement in quantitative traits using traditional strategies will be more time consuming. For examples, selection for parthenocarpy will be difficult using visual appearance as was done in the past. It is fairly straight forward to select cucumber genotypes containing high beta carotene from asegregating population, but it is difficult to distinguish levels of carotene based on color.

The most effective method for selecting a multiple allele trait is to utilize multiple markers to identify a majority of the alleles. This is especially true to the cucumber crop, which requires a large amount of space to evaluate. However, phenotypic markers will have an intrinsic disadvantage if the trait is influenced by the environmental factors. Development of disease resistance lines are often challenging for cucumber crop. Many disease resistance traits are quantitative; their expression is affected by environmental factors and requires complex inoculation procedure. Cucumber mosaic virus, downy mildew and powdery mildew are the good examples of diseases where development of new cultivar with high level of resistance has thus far proven difficult. Host plant resistance strategy should be utilized for control of these diseases. For these diseases several resistant sources have been identified [91] across the world but these genotypes do not withstand with high disease pressure under multiple locations.

3.5 Rationale for molecular breeding

Molecular markers have the potential to overcoming the limitations of traditional selection methods, since they are non-destructive, eliminate environmental variation associated with disease resistance and can be evaluated for multiple traits simultaneously. However, the molecular breeding requires the development of segregating populations for the traits of interest and the trait must be properly identified during marker identification. Recently, due to the advance technologies of genome sequencing, genome wide association study (GWAS) or Linkage disequilibrium (LD) mapping have gained popularity which is a powerful and alternative genetic mapping approach for the identification and dissecting important QTL regions which harbor candidate genes of interest in plants [92]. When, GWAS performed on large set of diversity panel, it provides the higher resolution mapping of traits associated variants because it exploits historical recombination events. GWAS study identifies genomic regions harboring loci controlling different traits. The sequencing data will help in the development of desirable genotypes in the form of cultivar which will help the farmers to receive higher returns on their investments. The potential of molecular breeding to save money is in their long term utilization in combination with multiple markers for a wide variety of traits; this will allow cucumber breeder to select for multiple traits from the large populations.

4. Conclusion

Classical genetics and traditional breeding techniques have made great contribution in better understanding and genetic improvement of cucumber crop for several qualitative and quantitative traits, understanding of phylogenetic relationship and taxonomy. Using traditional breeding methods, cucumber breeder have identified a number of genes associated with economic traits and used this information to develop early and high yielding cultivars. Advancement of some quantitative characters using traditional methodology is difficult and time consuming process. Therefore molecular markers technology offers an avenue to overcome the hurdles associated with traditional breeding. Molecular breeding has played a significant role in better understanding of cucumber genetics, and has been directly responsible for some of the improvement made in modern cucumber cultivars in different market classes. Classical breeders have said "Anything is possible using traditional approaches; it is just that the world is not large enough to hold the populations needed to find the variation required for some traits. As we move forward with molecular breeding in cucumber, it is important that we understand the need to maintain traditional breeding programs, and that the skill set required for classical breeding is not lost.

Conflict of interest

The authors declare no conflict of interest.

Author details

Gograj Singh Jat*, Tusar Kanti Behera, Suman Lata and Sachin Kumar
Division of Vegetable Science, ICAR-Indian Agricultural Research Institute, Pusa Campus, New Delhi, India

*Address all correspondence to: singhgograj@gmail.com

References

[1] Weng Y, Sun ZY. Major cucurbits (Chapter 1). In Y. H. Wang, T. K. Behera, & C. Kole (Ed.), Genetics, genomics and breeding of cucurbits (pp. 1-16). New York: CRC Press. 2012.

[2] Weng Y. Molecular Breeding Research in USDA-ARS Cucumber Improvement Program–Past, present and future. Journal of Tianjin Agricultural Science. 25(6) 6-18; 2019.

[3] Yang L, Koo DH, Li YH, Zhang XJ, Luan FS, Havey MJ, Jiang JM, Weng Y. Chromosome rearrangements during domestication of cucumber as revealed by high-density genetic mapping and draft genome assembly. Plant Journal 71: 895-906; 2012.

[4] Arakelyan H. Cucumber (*Cucumis sativus*)-1 Healthy Vegetables.–Mother Nature Healing 2019

[5] Akhtar P, Ahmad I, Jameela A, Ashfaque M, Begum Z. Energizing Effectiveness of Cucumber (Khayarain) For Health. A Review Article. Journal of Emerging Technologies and Innovative Research. 7(11); 2020.

[6] Dijkhuizen A, Kennard WC, Havey MJ, Staub JE. RFLP variability and genetic relationships in cultivated cucumber. Euphytica 90: 79-89; 1996.

[7] Kumar R, Munshi AD, Behera TK, Jat GS, Choudhary H, Singh M, Talukdar A. Genetic diversity of cucumber (*Cucumis sativus*) accessions differing in quantitative traits and microsatellite markers *Indian Journal of Agricultural Sciences* 90 (11): 2161-7; 2020.

[8] Miao H, Zhang S, Wang X, Zhang Z, Li M, Mu S., ... Gu X. A linkage map of cultivated cucumber (Cucumis sativus L.) with 248 microsatellite marker loci and seven genes for horticulturally important traits. Euphytica, 182, 167-176.2011; doi:10.1007/s10681-011-0410-5

[9] Weng Y, Colle M, Wang YH, Yang LM, Rubinstein M, Sherman A, Ophir R, Grumet R. QTL mapping in multiple populations and development stages reveals dynamic quantitative trait loci for fruit size in cucumbers of different market classes. Theoretical and Applied Genetics 128, 1747-1763; 2015

[10] Staub JE, Chung S-M, Fazio G. Conformity and genetic relatedness estimation in crop species having a narrow genetic base: The case of cucumber (Cucumis sativus L.). Plant Breeding, 124, 44-53; 2005. doi:10.1111/pbr.2005.124.issue-1

[11] Weng Y. Public sector cucumber research priority survey, Cucurbit Genetics Cooperative Report Issues 31-32:1-4; 2010.

[12] Behera TK, Staub JE, Behera S, Delannay IY, Chen JF. Marker-assisted backcross selection in an interspecific Cucumis population broadens the genetic base of cucumber (*Cucumis sativus* L.). Euphytica. 2011; 178, 261-272. doi:10.1007/s10681-010-0315-8

[13] Farag MI, Behera TK, Munshi AD, Bharadwaj C, Jat GS, Khanna M, Chinnusamy V. Physiological analysis of drought tolerance of cucumber (Cucumis sativus L.) genotypes under three soil moisture-deficit regimes in field conditions. Indian Journal of Agricultural Sciences 89 (9): 1445-1450. 2019

[14] Munshi AD, Tomar BS, Jat GS and Singh J (2017). Quality seed production of open pollinated varieties and F 1 hybrids in cucurbitaceous vegetables. In: Compendium-ICAR Sponsored Ten days short course on "Advances in variety maintenance and quality seed production for entrepreneurship" February 14-23 2017, pp 107-125.

[15] Huang, S.W., Li, R.Q., Zhang, Z.H., Li, L., Gu, X.F., & Fan, W. (2009). The genome of the cucumber, *Cucumis sativus* L. Nature Genetics, 41, 1275-1281. doi:10.1038/ng.475

[16] Weng YQ, Johnson S, Staub JE, Huang SW. An extended microsatellite genetic map of cucumber (*Cucumis sativus* L). HortScience, 45, 880-886; 2010.

[17] Zhang SP, Miao H, Gu XF, Yang YH, Xie BY, Wang XW, Sun RF. Genetic mapping of the scab resistance gene Ccu in cucumber. Journal of the American Society of Horticulture Science, 135, 53-58; 2010.

[18] Zhang WW, He H, Yuan G, Du H, Yuan LH, Li Z, Cai R. Identification and mapping of molecular markers linked to the tuberculate fruit gene in the cucumber (*Cucumis sativus* L.). Theoretical and Applied Genetics, 120, 645-654; 2009. doi:10.1007/s00122-009-1182-3

[19] Sheng Y, Pan Y, Li Y, Yang L, Weng Y Quantitative trait loci for fruit size and flowering time-related traits under domestication and diversifying selection in cucumber (*Cucumis sativus*) 2019. DOI: 10.1111/pbr.12754

[20] Wenzel G, Kennard WC, Havey MJ. Quantitative trait analysis of fruit quality in cucumber: *The Journal of Horticultural Science and Biotechnology* 31 QTL detection, confirmation, and comparison with mating- design variation. Theoretical and Applied Genetics, 91, 53-61; 1995. doi:10.1007/BF00220858

[21] Park Y, Sensoy S, Wye C, Antonise R, Peleman J, Havey MJ. A genetic map of cucumber composed of RAPDs, RFLPs, AFLPs, and loci conditioning resistance to papaya ring spot and zucchini yellow mosaic viruses. Genome, 43, 1003-1010; 2000. doi:10.1139/g00-075

[22] Fazio G, Staub JE, Stevens MR. Genetic mapping and QTL analysis of horticultural traits in cucumber (*Cucumis sativus* L.) using recombinant inbred lines. Theoretical and Applied Genetics, 107, 864-874. 2003; doi:10.1007/s00122-003-1277-1

[23] Zhang S, Liu S, Miao H, Wang M, Liu P, Wehner TC, Gu X. Molecular mapping and candidate gene analysis for numerous spines on the fruit of Cucumber. Journal of Heredity 107, 471-477; 2016. doi:10.1093/jhered/esw028

[24] Zhang S, Liu MM, Miao H et al. Chromosomal mapping and QTL analysis of resistance to downy mildew in *Cucumis sativus*. Plant Disease 97:245-251; 2013a.

[25] Zhang S, Miao H, Sun R, Wang X, Huang S, Wehner TC, Gu X. Localization of a New Gene for Bitterness in Cucumber. Journal of Heredity 2013b:104(1):134-139 doi:10.1093/jhered/ess075

[26] Zhang SP, Liu MM, Miao H, Zhang SQ, Yang YH, Xie BY, Gu XF. Chromosomal mapping and QTL analysis of resistance to downy mildew in Cucumis sativus. Plant Disease, 97, 245-251; 2013c. doi:10.1094/PDIS-11-11-0941-RE

[27] Lu HW, Miao H, Tian GL, Wehner TC, Gu XF, Zhang SP. Molecular mapping and candidate gene analysis for yellow fruit flesh in cucumber. Molecular Breeding, 35, 64. doi:10.1007/s11032-015-0263-z; 2015.

[28] Pan Y, Wen C, Han Y et al. QTL for horticulturally important traits associated with pleiotropic andromonoecy and carpel number loci, and a paracentric inversion in cucumber. Theor Appl Genet 133, 2271-2290 (2020a). doi:10.1007/s00122-020-03596-y

[29] Pan Y, Wen C, Han Y, Wang Y, Li Y, Li S, Cheng X, Weng Y. QTL for horticulturally important traits associated with pleiotropic andromonoecy and carpel number loci, and a paracentric inversion in cucumber Theoretical and Applied Genetics 133(7); 2020b. DOI: 10.1007/s00122-020-03596-y

[30] Wang Y, Jiang B, Dymerski R et al Quantitative trait loci for horticulturally important traits defining the Sikkim cucumber, *Cucumis sativus* var. *sikkimensis*. *Theor Appl Genet* 134, 229-247; 2021. doi:10.1007/s00122-020-03693-y

[31] Pierce LK, Wehner TC. Review of genes and linkage groups in cucumber. HortScience, 25, 605-615; 1990.

[32] Knopf RR, Trebitsh T. The female-specific Cs-ACS1G gene of cucumber. A case of gene duplication and recombination between the non-sex-specific 1-aminocyclopropane-1-carboxylate synthase gene and a branched-chain amino acid transaminase gene. Plant Cell Physiol 47:1217-1228; 2006

[33] More TA, Munger HA. Effect of temperature and photoperiod on gynoecious sex expression and stability in cucumber. Vegetable Science, 14, 42-50; 1987.

[34] Perl-Treves R, Rajagopalan PA. Close, yet separate: Patterns of male and female floral development in monecious species. In C. Ainsworth (Ed.), Flower development and manipulation (pp. 117-146). Oxford: Blackwell; 2006.

[35] Li J, Zhiwei Q, Xiuyan Z. Genetic analysis of gynoecious in cucumber (*Cucumis sativus L.*). Institute of Agricultural Information, Chinese Academy of Agricultural Sciences. http://www.caas.net.cn. 2012

[36] Lou FQ, Chen FJ, Chen ZL, Wolukau NJ. Identification of an AFLP marker linked to a locus controlling gynoecy in cucumber and its conversion into SCAR marker useful for plant breeding. Acta Horticulturae. 763: 75-82; 2007

[37] Mibus H, Tatlioglu T. Molecular characterization and isolation of the F/f gene for femaleness in cucumber (Cucumis sativus L.). Theoretical and Applied Genetics, 109, 1669-1676. 2004; doi:10.1007/s00122-004-1793-7

[38] Jat GS, Munshi AD, Behera TK, Choudhary H, Dash P, Amarnath R, Kumari S. Genetics and molecular mapping of gynoecious (F) locus in cucumber (*Cucumis sativus* L.) Journal of Horticultural Science and Biotechnology. 94:1, 24-32. 2018

[39] Pati K, Munshi AD, Behera TK. Inheritance of gynoecism in Cucumber (*Cucumis sativus* L.) using genotype GBS-1 as gynoecious parent. Genetika, 47, 349-356; 2015. doi:10.2298/GENSR1501349P

[40] Shengjun Z, Peng Z, Yuquing Z, Xinjuan C, Liping C. Identification of SSR markers linked to gynoecious loci in cucumber (*Cucumis sativus* L). Journal of Zhejiang University (Agril. & Life Sci.), 39, 291-298; 2013.

[41] Win KT, Zhang C, Silva RR, Lee JH, Kim Y-C, Lee S. Identification of quantitative trait loci governing subgynoecy in cucumber Theor Appl Genet. 2019 May;132(5):1505-1521; doi: 10.1007/s00122-019-03295-3.

[42] Boopalakrishnan G. Jayavel S, Behera TK, Munshi AD, Choudhary H, Singh A, Pandey MK. Mapping and linkage analysis of epistatic QTLs for gynoecious trait (F) in cucumber (*Cucumis sativus* L.), The Journal of Horticultural Science and Biotechnology, 2020; DOI: 10.1080/14620316.2020.1855084

[43] Pike LM, Peterson CE. Inheritance of parthenocarpy in the cucumber

(Cucumis sativus L.).Euphytica, 18: 101-105; 1969.

[44] Kim S, Okubo H, Fujieda K. Endogenous levels of IAA in relation to parthenocarpy in cucumber (*Cucumis sativus* L.). Scientia Horticulturae, 52: 1-8. 1992

[45] Juldasheva LM. Inheritance of the tendency towards parthenocarpy in cucumbers. Vavilova, 32: 58-59; 1973.

[46] Sun Z, Lower RL, Staub JE. Analysis of generation means and components of variance for parthenocarpy in cucumber (*Cucumis sativus* L.). Plant Breed., 123(3): 277-280; 2006a

[47] Sun Z, Lower RL, Staub JE Variance component analysis of parthenocarpy in elite U.S. processing type cucumber (*Cucumis sativus* L.) lines. Euphytica, 148(3): 331-339; 2006b.

[48] Lietzow CD, Zhu H, Pandey S et al. QTL mapping of parthenocarpic fruit set in North American processing cucumber. Theor Appl Genet 129, 2387-2401 (2016). doi:10.1007/s00122-016-2778-z

[49] Wu Z, Zhang T, Li L et al. Identification of a stable major-effect QTL (Parth 2.1) controlling parthenocarpy in cucumber and associated candidate gene analysis via whole genome re-sequencing. BMC Plant Biol 16, 182; 2016a. doi:10.1186/s12870-016-0873-6

[50] Wu Z, Zhang T, Li L, Xu J, Qin X, Zhang T, Cui L, Lou Q, Li J. Identification of a stable major-effect QTL (Parth 2.1) controlling parthenocarpy in cucumber and associated candidate gene analysis via whole genome resequencing. BMC Plant Biol., 16(1):182; 2016b.

[51] Yuan XJ, Pan JS, Cai R, Guan Y, Lui LZ et al. Genetic mapping and QTL analysis of fruit and flower related traits in cucumber (*Cucumis sativus* L.) using recombinant inbred lines. Euphytica 164:473-491; 2008

[52] Pramnoi P, Somta P, Chankaew S, Juwattanasomran R Srinives P. A single recessive gene controls fragrance in cucumber (*Cucumis sativus* L.) J. Genet. 92, 147-149; 2013.

[53] Liu H, Meng H, Pan Y, Liang X, Jiao J, Li Y, Chen S, Cheng Z. Fine genetic mapping of the white immature fruit color gene w to a 33.0-kb region in cucumber (*Cucumis sativus* L.) Theoretical and Applied Genetics 2015; DOI 10.1007/s00122-015-2592-z

[54] Bo K, Song H, Shen J, Qian C, Staub JE, Simon PW, Lou Q, Chen J. Inheritance and mapping of the ore gene controlling the quantity of b-carotene in cucumber (*Cucumis sativus* L.) endocarp Mol Breeding. 2011. DOI 10.1007/s11032-011-9624-4

[55] Bo K, Wei S, Wang W et al. QTL mapping and genome-wide association study reveal two novel loci associated with green flesh color in cucumber. BMC Plant Biol 19, 243. 2019; https://doi.org/10.1186/s12870-019-1835-6

[56] Zhang P, Qin ZW, Wang LL, Zhou XY, Wang GL. Mapping quantitative traits loci linked to bending fruit trait in cucumber. Journal of Northeast Agricultural University (English Edition), 18, 1-5; 2011.

[57] Klosinska U, Kozik EU. Inheritance of a New Trait—Twin Fused Fruit—in Cucumber Hortscience 41(2):313-314; 2006.

[58] Wang Y, Bo K, Gu X, Pan J, Li Y, Chen J, Wen C, Ren Z, Ren H, Chen X, Grumet R and Weng Y. Molecularly tagged genes and quantitative trait loci in cucumber with recommendations for QTL nomenclature. Horticulture Research 7:3 2020. doi:10.1038/s41438-019-0226-3

[59] McGrath MT Fungicide resistance in cucurbit powdery mildew: experiences and challenges. Plant Dis 85:236-245; 2001.

[60] de Ruiter W, Hofstede R, de Vries J, van den HH. Combining QTL for resistance to CYSDV and powdery mildew in a single cucumber line. In: Pitrat M (ed) Proceedings of IXth EUCARPIA meeting on genetics and breeding of Cucurbitaceae. INRA, Avignon, pp. 181-188; 2008.

[61] Liu L, Yuan X, Cai R, Pan J, Huanle H, Yuan L, Yuan G, Zhu I. Quantitative trait loci for resistance to powdery mildew in cucumber under seedling spray inoculation and leaf disc infection. J Phytopathol 156:691-697; 2008.

[62] Xu Q, Shi Y, Yu T, Xu X, Yan Y, Qi X, Chen X Whole-genome resequencing of a cucumber chromosome segment substitution line and its recurrent parent to identify candidate genes governing powdery mildew resistance. PLoS ONE 11:e0164469; 2016

[63] Smith PG. Powdery mildew resistance in cucumber. Phytopathology 39:1027-1028; 1948.

[64] Barnes WC, Epps WM. Powdery mildew resistance in South Carolina cucumbers. Plant Disease Report. 1956; 40:1093

[65] Kooistra E (1968) Powdery mildew resistance in cucumber. Euphytica 17:236-244

[66] Shanmugasundaram S, Williams PH, Peterson CE. A recessive cotyledon marker gene in cucumber with pleiotropic effects. HortScience 7:555-556; 1972.

[67] Shanmugasundaram S, Williams PH, Peterson CE. Inheritance of resistance to powdery mildew in cucumber. Phytopathology 61:1218-1221; 1971.

[68] Fanourakis NE, Simon PW. Analysis of genetic linkage in the cucumber. J. Heredity 78:238-242. 1987

[69] Van Vliet, G.J.A. and W.D. Meysing. 1977. Relation in the inheritance of resistance to *Pseudoperonospora cubensis* Rost and Sphaerotheca fuliginea Poll. in cucumber (*Cucumis sativus* L.). Euphytica 26:793-796.

[70] Badr LAA, Mohamed FG. Inheritance and nature of resistance to downy mildew disease in cucumber (*Cucumis sativus* L.). Annals of Agriculture Science, Moshtohor 1998; 36(4):2517-2544.

[71] Angelov D. Inheritance of resistance to downy mildew, *Pseudoperonospora cubensis* (Berk. & Curt.) Rostow. Rep. 2nd National Symposium Plant Immunity (Plovdiv). 1994; 3:99-105.

[72] Doruchowski RW Lakowska-Ryk E. Inheritance of resistance to downy mildew (Pseudoperonospora cubensis Berk & Curt) in *Cucumis sativus*. Proc. 5th EUCARPIA Symp, 27-31 July, Warsaw, Poland. (R. W. Doruchowski, E. Kozik, and K. NiemirowiczSzczytt, eds.), pp. 132-138. Res. Inst. Veg. Crops, and Warsaw Univ. Agric., Warsaw, Poland. 1992.

[73] Petrov L, Boodert K, Sheck L, Baider A, Rubin E, Cohen Y, Katzir N, Paris HS. Resistance to downy mildew, Pseudoperonospora cubensis, in cucumbers. Acta Horticulturae 510:203-209; 2000.

[74] Kozik EU, Klosin'ska U, Call AD, Wehner TC. Heritability and genetic variance estimates for resistance to downy mildew in cucumber accession Ames 2354. Crop Sci 53:177-182 2013.

[75] Chen T, Katz D, Ben NY, Hammer R, Ben Daniel BH, Rubin AE, Cohen Y. Isolate-Dependent Inheritance of Resistance Against Pseudoperonospora cubensis in Cucumber. Agronomy 2020,

10, 1086. doi:10.3390/agronomy10081086

[76] Innark P, Panyanitikoon H, Khanobdee C, Samipak S, Jantasuriyarat C. QTL identification for downy mildew resistance in cucumber using genetic linkage map based on SSR markers Journal of Genetics (2020) 99:81; 2020. doi:10.1007/s12041-020-01242-6

[77] Shi L et al. Inheritance and QTL mapping of cucumber mosaic virus resistance in cucumber (*Cucumis Sativus* L.). PLoS ONE 13, e0200571 (2018).

[78] Munshi AD, Panda B, Mandal B, Bisht IS, Rao ES, Kumar R. Genetics of resistance to Cucumber mosaic virus in *Cucumis sativus* var. *hardwickii* Euphytica. 2008; 164:501-507 DOI 10.1007/s10681-008-9741-2

[79] Tian G et al. Genetic analysis and gene mapping of papaya ring spot virus resistance in cucumber. Mol. Breed. 35, 110 (2015).

[80] Tian G. et al. Genetic analysis and fine mapping of *Watermelon mosaic virus* resistance gene in cucumber. Mol. Breed. **36**, 1-11 (2016).

[81] Kumar, A., Bhattarai, A., Rathore, A.S., Jat, G.S., Behera, T. K., Renukadevi, P., Baranwal, V. K., Jain R. K. and Basavaraj Y. B. (2021). Association of cucurbit aphid-borne yellows virus with cucumber plants in India. VirusDis doi:10.1007/s13337-020-00645-4

[82] Airina CK, Pradeepkumar T, George TE, Sadhankumar PG, Krishnan S Heterosis breeding exploiting gynoecy in cucumber (*Cucumis sativus* L.). Journal of Tropical Agriculture 2013; 51: 144-148.

[83] Jat GS, Munshi AD, Behera TK, Choudhary H, Dev B Exploitation of heterosis in cucumber for earliness, yield and yield components utilizing gynoecious lines. Indian Journal of Horticulture 72 (4): 494-499; 2015

[84] Pandey S, Singh B, Singh M Rai M. Heterosis in cucumber (*Cucumis sativus* L.). Veg. Sci. 32: 143-145; 2005.

[85] Jat GS, Munshi AD, Behera TK, Bhardwaj C. Inheritance of parthenocarpy in gynoecious cucumber (*Cucumis sativus* L.) cultivar PPC-2 Journal of Horticultural Science. 12(2): 193-197; 2017.

[86] Kumar S, Kumar R, Kumar D, Gautam N, Singh N, Parkash C, Dhiman MS, Shukla YR. Heterotic potential, potence ratio, combining ability and genetic control of yield and its contributing traits in cucumber (*Cucumis sativus* L.), New Zealand Journal of Crop and Horticultural Science, 45:3, 175-190; 2017. DOI: 10.1080/01140671.2016.1270336

[87] Preethi GP, Anjanappa M, Ramachandra RK, Vishnuvardhana. Heterosis Studies for Growth and Yield Traits in Cucumber (*Cucumis sativus* L.) Int. J. Curr. Microbiol. App. Sci. 8(3):916924; 2019. doi:https://doi.org/10.20546/ijcmas.2019.803.109

[88] Simi F, Ivy N, Saif H, Akter S, Anik M. Heterosis in cucumber (*Cucumis sativus* L.). *Bangladesh* Journal of Agricultural Research, 42(4), 731-747; 2018. doi:10.3329/bjar.v42i4.35801

[89] Singh R, Singh AK, Kumar S, Singh BK. Heterosis and inbreeding depression for fruit characters in cucumber Indian J. Hort. 69: 200-204; 2012.

[90] Jat GS, Munshi AD, Behera TK, Tomar BS. Combining ability estimation of gynoecious and monoecious hybrids for yield and earliness in cucumber (Cucumis sativus L.) Indian Journal of Agricultural Sciences 86 (3): 399-403. 2016

[91] Gautam D, Nath R, Gaikwad AB, Bhat KV, Mondal B, Akhtar J, Jat GS, Iquebal AM, Tiwari B, Archak S. Identification of new resistant sources against downy mildew disease from a selected set of cucumber germplasm and its wild relatives. Indian Journal of Genetics and Plant Breeding 8 (4): 427-431. 2020

[92] Zhu C, Gore M, Buckler ES, Yu J. Status and prospects of association mapping in plants. Plant Genome 1: 5-20; 2008.